Dedication

For Frank Amato
With thanks for help, encouragement,
and risk taking.

Photo credit to J.L. Schollmeyer
p. 14, 77, 115, 120, 135, 141.

ISBN 0-936608-59-5
Copyright 1987 by Dave Hughes
Printed in U.S.A.

Typesetting: Chris Mazzuca   Book Design: Joyce Herbst

Western
Streamside
Guide

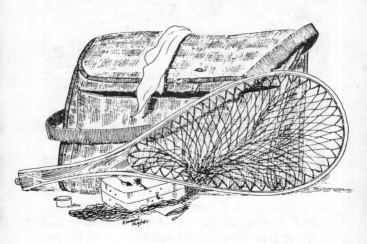

Dave Hughes

Frank Amato Publications
P.O. Box 02112
Portland, Oregon 97202

About the Author

Dave Hughes is a professional writer and amateur aquatic entomologist. He has spent more than twenty years fly fishing over insect hatches on Western waters. He is a regular contributor to *Flyfishing, Fly Fisherman,* and *Salmon Trout Steelheader* magazines. His feature articles have also appeared in *Fly Fisher, Fly Tyer, Rod & Reel, Outdoor Life,* and *Field & Stream.*

Dave is a charter member of the Rainland Fly Casters, was founding president of Oregon Trout, and received the 1985 Lew Jewett Memorial Life Membership from the Federation of Flyfishers. His other books include *Western Hatches,* co-authored with Rick Hafele, *An Angler's Astoria, American Fly Tying Manual, Handbook of Hatches,* and *Flyfishing* magazine's *Western Fly Fishing Guide.*

Hughes lives in his hometown, Astoria, Oregon.

Contents

Observation and Imitation

Chapter 1 _____

Art Flick's classic little *Streamside Guide* was first published in 1947. Forty years later it is still a valuable book because it details the few Eastern hatches that an angler is most likely to encounter.

A streamside guide for western waters can be just as valuable, but it can't focus quite so narrowly. Mr. Flick's book dealt primarily with mayfly duns. Just a few pages were devoted to nymphs, and to stoneflies and caddisflies. Even in the East it is now conceded that mayflies are not the only important order of insects, and the adults are not the only important stage.

In order to be as useful as Flick's Eastern streamside guide, a *Western Streamside Guide* must have a wider scope. And it must poke its nose under water a time or two, holding its breath while peering at the sub-aquatic stages which trout devour with such delightful regularity.

The wonderful and confusing geography of the West has led to the evolution of a wonderful and confusing array of aquatic insect species. Most of these suffer moments of vulnerability to trout. When they do they are the most important insects around, to the trout fisherman trying to match them. But for many species these moments of most importance are limited to specific situations, brief in time, or isolated to a few rivers or lakes.

Only a few species of insects are widespread, important nearly everywhere in the West. These are the hatches that you are sure to encounter as a Western angler, whether you fish out the entire season on your home waters, or pack your travel rod to visit the various meccas of Western angling. *This book is about those few species that are going to frustrate you at one time or another while fishing through the Western season, no matter where you fish. It is about the insects, the flies that match them, and the tactics that fool the trout that feed on them.*

There are important differences between East and West. On our varied Western waters, hatches tend to fall into groupings of species, rather than springing from a single species. For instance the Eastern Green Drake is an *Ephemera guttulata*. The Western Green Drake can be one of three species: *Ephemerella grandis, E. doddsi,* or *E. glacialis.* They are so similar that a single pattern works for all three.

Emergence periods are fairly clearly defined in the East. In the West, elevation changes cause some hatches to unfold over a period of months rather than weeks. For instance the famous salmonfly, which emerges in April in the sea-level lowlands of western Oregon, fails to hatch in exciting numbers until July on some high-elevation streams in Montana.

Our understanding of hatches in the West is still being pieced together. There has been much jumping to conclusions, especially in magazine articles. Some stuff that has been written has been wrong. Some of it has been mine. Though it is regrettable that a lot of the speculation has been done in

7

print, it is also likely that our youthful Western impetuousness has led us to it, and that we have learned a lot from it. The trail to the truth is made more tortuous by wrongful speculation, but it is also often sped by it.

Another major difference between East and West is the number of species that are important in specific instances without being widespread enough to be encountered often. I have a favorite desert river that I fish every fall. I know the hatches I will encounter then, and know with some certainty what flies I can carry to catch trout. Lots of trout. But I fished it once in early spring, and got a lesson, the outcome of which I hate to confess.

It was April. The air was warm. The sage hills were green and struck by the sun, the first time I had seen this river surrounded by anything but sad browns and grays. Pairs of mallards and Canada geese erupted scoldingly from backwaters as I wandered from riffle to pool along the river. Clever killdeer led me away with broken wings, circling back behind me to reveal unerringly the location of their eggs nested out on the open stones.

About mid-morning a hatch began. They were mayfly duns, size 14, with olive bodies and blue wings. This was not a size and color combination that this river produced in fall, and it was therefore not a fly pattern that I carried in my fly boxes for this river. And like a fool, those were the only fly boxes I'd brought on the trip.

The duns emerged in a pool, in the first few feet of calm water below a wide riffle that was only inches deep. Trout formed a skirmish line across the head of the pool, feeding greedily on the insects as quickly as the helpless duns washed down from the riffle above. A few duns escaped into the air simply because so many hatched the trout couldn't eat them all. It was a busy little intersection.

I took the time to collect a specimen, and compared it to the flies in my box, to confirm that I could not match it. But I wasn't worried. Everybody knows desert trout are dumb. Selectivity in a place like this would be time wasted by the trout. I cast my eyes up and down the open banks of the river. Not only were there no other fishermen in sight, there was no evidence that any other fishermen had been on the river all spring. These fish had not even been cast over.

I tied on a size 14 Adams, tested the tippet knot, and prepared to catch a bunch of trout. But it wasn't the trout that wasted their time. Two hours and a dozen frustrated fly patterns later, I reeled in and waded into the riffle to get a closer look at what sort of critter the river was delivering to its trout. I scooped up a dun in my hand and examined it closely. It could have been one of several things, because duns are hard to identify. But the nymphs that cascaded into my collecting net, when I held it suspended in the current, were flat, three-tailed and unmistakably *Rhithrogena*.

I finally fooled a few fish. But it was damned few, and I did it with a wet fly that looked a bit like the emerging nymph, not a dry fly that looked like the emerged dun. It was an embarrassing lesson to learn about the selectivity of desert trout. As I glanced again up and down the deserted river, I was glad that nobody was around to witness my humbling.

I've never gotten back to that river in spring, and I've never encountered that blue-winged olive *Rhithrogena* again.

It is hard to define important insects as opposed to unimportant ones. When an insect is suddenly

9

abundant, and suddenly the only thing that interests trout, it is also suddenly very important. But you can clutter up your fly boxes pretty fast with patterns that match insects you've seen only once, and might never see again.

When I first got interested in aquatic insects and their intimate relationships with trout, I invented entire batteries of fly patterns for every insect I collected, conjured up abbreviated lists for those that I glimpsed as they winged past. I soon had my fly boxes, then my entire vest, so bulging with the useless that I could not find the useful when I needed it.

That wasn't a practical approach to constructing a useful set of western trout flies. Sorting out the important hatches from the unimportant hatches is a practical approach.

Which insects are important? They are the species most widespread geographically. They are the species that live in *trout water* as opposed to those that live in marginal habitats where trout are not happy to be found. They are the species that hatch in ways that make them vulnerable to trout, in concentrations that induce trout to feed on them selectively. They are the ones you and I are most likely to encounter being fed on by trout.

Fly boxes containing patterns that match the Western species with the above characteristics will be fly boxes filled with useful flies. If you are an active fisherman, and get out fishing a lot, you will use almost all of the patterns at one time or another during the course of a season or two. You will also come up short a time or two. The boxes won't contain every fly you will need.

I'm sure that my opinion about the most important Western hatches will start some arguments, even among my friends. But this book is based on my own experience, and I do have certain prejudices.

I prefer to fish streams over lakes, though lake insects are not as varied as stream insects, so I think I've fished most of the important stillwater hatches. I prefer to fish dry flies, wets, or unweighted nymphs over weighted nymphs. I've included many nymph dressings in this book, but I know I need to learn a lot more about them. I know it because of the number of times experts like Rick Hafele, who co-authored *Western Hatches* with me, have given me embarrassing lessons by fishing nymphs along the bottom while I fished something else up toward the surface.

I prefer quick rods over slow ones. But that tells you more about me than it does about fishing, so let's get back to the subject. I prefer flies that fish successfully on a wide variety of water types over flies that fish only on the very quietest waters or the very fastest. That is why I choose a parachute tie for the Little Olive, rather than a no-hackle, which is preferred by most fishing writers. The no-hackle simply won't float on the fairly fast waters where I encounter Little Olives most often. A parachute will take trout there, and also when I encounter the small *Baetis* duns boating spring creek currents.

My experience is more on the general run of Western freestone streams, less on the storied spring creeks. I fish home waters more often than I travel. And I find, as a rule, that it is better if my flies float where I cast them than it is if they float where I wish I could cast them.

I am a rough-and-ready fly tier. My flies tend to be simple, easily tied out of readily-obtainable materials. I believe that presentation is at least as important as pattern. I fished with Richard Bunse once on a spring creek that is not too far from my home waters. What we learned that day drilled the importance of presentation into a couple of relatively resistant heads.

Bunse is a remarkable dry fly fisherman, the best I have met. He is also an excellent tier, one of those astonishing types who not only has Green Drake dressings in his fly box when he goes to fish the Green Drake hatch, but also has three or four different experimental patterns to try on the trout, to see which they might prefer. It was a Green Drake hatch we went to fish on this spring creek. We were not disappointed.

Drakes emerged all around us, and trout rose to take them eagerly. But we weren't catching any fish. Bunse and I both ran through his entire corps of Green Drake dressings. We cast them upstream on short lines, landing them delicately just a foot or two above the rising fish. We used curve casts and hook casts and all kinds of casts to get the flies over the fish without spooking them. And we succeeded. We didn't spook the fish. But we didn't catch them, either. They kept rising, but not to our flies.

Then Bunse did a thing which was radical at the time he did it, which was quite a few years ago. He turned around, established the rise rhythm of a trout feeding below him, then cast downstream to it. He let his line land on the water with lots of slack in it. As this slack line fed out, the fly drifted primly down over the fish, arriving ahead of the line and leader. The fish rose to take it confidently.

After that we both turned around and fished downstream. And we both began catching fish. We tried once again all of the Green Drake dressings the fish had previously rejected. And we found that the fish would take them all. Pattern didn't have much to do with our problem.

The downstream slack-line cast is now elemental information in every article and book about fly fishing. But back then it was an early lesson in the importance of presentation and fly pattern. You've got to get both right if you want to fool fish.

Observation, such as the kind Bunse made when he intuitively turned around and fished downstream, is one of the keys to successful fishing. Sometimes a small change in what you are doing can make a big change in your luck. Sometimes observation will tell you that trout are not doing what you think they are doing. For instance, it is easy to conclude that trout take adult caddis when lots of those bouncy fliers bat around in the air. But trout feed more often on ascending caddis pupae. A wet fly fished just beneath the surface will often take many more fish than a dry fly floated on the surface.

Observing and understanding the life cycles of aquatic insects will help you take trout. For example, free-living larval caddis, called Green Rock Worms, show up frequently in riffle net samples. But cranefly larvae often outnumber them in the same sample. These are bigger and juicier. If you were a trout, you would rather eat cranefly larvae. The cranefly larvae, however, live down in the gravel beneath the stream bed, while the caddis larvae prowl around the crevices in rocks up above the bed. Caddis worms are knocked loose by the current more often because of their active life style. Trout feed on them heavily, and an imitation of the caddis is likely to take trout in many more situations than an imitation of the cranefly.

In the West, with its wide array of aquatic insects, it is common to encounter a hatch that is not one of the major species, but looks so much like it that telling them apart is difficult. It is a little-known secret that trout also have difficulty telling them apart; they get confused counting the tails, reading the gill forms, keying the insects out by studying the pattern of the veins in their wings.

Trout have no need to key an insect out before eating it. You don't have to identify an insect to species before matching it, either. If it looks so much like a popular species that you can't tell them apart, then merely use the same dressing you would use for the more common hatch, and no trout will know the difference.

Imitation has its options. Different kinds of water call for different pattern styles. The dry fly that works best on the slick surface of a meadow stream might not even float on a tumbling plunge-pool stream. A fly that fools trout in fast water might frighten them on heavily fished water like the Henry's Fork of the Snake River.

Throughout this book I have keyed pattern selection to the preferred habitat of the natural. But I have also given options, so that hatches of the same insect encountered on other water types can be covered with other pattern styles. Where an insect lives in more than one type of habitat, I have selected a primary pattern that fishes the widest variety of water types. But if you fish only one type of water, you should choose the pattern style that fishes best on that kind of water.

The purpose of this *Western Streamside Guide* is not to tell you precisely where and when you will find each hatch. It is, rather, to prepare you to fish over the most common hatches in the West, where and when you find them. These are hatches you will almost surely encounter in the course of a season. If you are prepared for them, you will seldom get caught short of a pattern that will take fish, anywhere throughout the West, anytime during the season.

Western March Brown

Rhithrogena Morrisoni, hageni

Chapter 2

It was a typical torn March day. Most of the morning it rained. Then the sun broke out, and we ate lunch with high hopes for an afternoon hatch. But it began to hail before our sandwiches were gone, and with the hail we thought our hopes were gone. But before the last ring from a falling hailstone faded on the river, the first ring of a rising trout wrinkled its surface.

Rick Hafele said, "I think the hatch is on."

Tony Robnett and I nearly tipped over the drift boat getting at our fly rods. March Browns are the first major hatch of the Western season; if you don't get excited about them, your year is going to be a dull one.

It was a large western Oregon river. The three of us spread out comfortably along the edge of a long current tongue, where the water from a vigorous riffle above narrowed and slowed into a hundred-yard run below. March Brown duns began popping to the surface right along the seam of the current, where the fast water eddied back a bit to check on itself. Trout began to rise all along the seam.

Tony did very well below me. Every time I

looked down his way I saw rainbows, 12 to 14 inches long, dancing around at the end of his line. Rick did even better above. He landed and released a trout on nearly every cast. But the man in the middle wasn't doing so well.

"Oh, well," I thought, "I'll just call down and find out what Tony is using." I called down and asked him.

"March Brown Comparadun," he called back. "Same thing you should be using. You gave this one to me." He was right on both counts: it was the same thing I was using, and I had given his to him.

"Oh, hell!" I thought. "Why isn't mine working?" I considered asking Tony to give the fly back, but the scarce dignity I own wouldn't permit me to do it. I went back to my fishing, and soon discovered my problem.

I raised my rod to make another cast, and found that my fly was stuck in a trout. I shouted to make sure Rick and Tony knew I had one on, then played it out in the limelight as if it was not an accident. I caught a couple more that way, but the truth had already occurred to me: I couldn't see my fly, and I couldn't tell when a trout took it. The water I fished was more rumpled than Tony's, down below me.

I had some experimental patterns carelessly tossed into a Sucrets box that day. They were parachute ties, with high-rise wings standing above a whirring helicopter blade of stiff brown hackle. The dressings were based on a couple sent to me by Robert Borden, manufacturer of Hareline Dubbin'. "Try these when the March Browns hatch," Bob had told me. "The body rides down in the water, where the trout can see it. The wing rides up high, where you can see it."

Bob was right. Trout took the fly just as eagerly as they took the low-floating comparadun. But when they did it, I could see it happen. The fly was visible as it rode down along the edge of that slightly riffled current tongue. And the fly worked just as well when I moved upstream to fish the quiet water of the flat tailout, just above the riffle. Soon I was catching just as many fish as Tony, and a little dignity had seeped back into me. But Rick was another problem. I'll leave him standing at the foot of his riffle, and deal with him later.

March Browns begin to emerge in February. The hatch gets stronger in March, peaking toward the end of that month of confusing weather, but continuing strongly into the first weeks of April. One of the happiest things about this early hatch is its tendency to peak well before snow melt knocks Western rivers out of shape.

March Browns taper off in early May in the lower elevations, but hatches in the Rockies are just getting going then. I have fished over them on the Henry's Fork of the Snake, in Idaho, in early June. Hatches of closely related species go on into July and August. It is not uncommon to encounter a trailing, but substantial, hatch of March Browns in September on some Oregon rivers.

Spring and early summer hatches come off during the warmest part of the day. After a warm morning they might start at 11 o'clock and last until 2 or 3 o'clock. On a cold day they might hold off until 1 or even 2 o'clock, and hatch until 3 or 4 o'clock. Days that are cloudy but mild in temperature are best: the hatch will sometimes last four or five hours. When the sun is out and bright, the hatch can be compacted into a frenzied hour or less.

March Browns are reported in collection samples along the coast from California north to British Columbia, and inland all the way from New Mexico to Alberta, including all of the Rocky Mountain states. The March Brown, or closely related species that can be fished with the same flies, inhabit Oregon's Deschutes, Montana's Madison and Rock Creek, and Colorado's Frying Pan, along with most other heavily riffled rivers.

Nymphs of the *Rhithrogena* live in fast and highly oxygenated currents. They cling to stones in the swiftest water, and are abundant in most of the freestone rivers throughout the Western states. They would be a better-known hatch if more people were willing to submerge their timid toes into frigid pre-runoff waters.

March Brown nymphs belong to the *clinger* category of mayflies. They are flattened, an adaptation to living in the skinny layer of water on a rock where friction slows a rushing current. They hold on tenaciously, and are not available in such numbers that trout are likely to feed selectively on them.

The nymphs begin to migrate out of fast water prior to emergence. Their habitat during most of the one-year life cycle is almost brutal. But by the time they are ready to take a chance on ephemeral flight they are most abundant in the quiet water above, below, or off to the side of tumbled water. The best places to fish a March Brown hatch are almost always in the nearest semi-quiet water to water that is moderate to swift.

At emergence the nymphs swim rather awkwardly but fairly quickly to the surface, flipping their abdomens up and down with an effort that shoud gain them greater speed than it does. They are vulnerable to feeding fish during this time of troubled transition to the surface.

The nymphal shuck splits along the back in the surface film or just below it. The dun works its way out and leaves the water as soon as it can. With the varied weather of spring, this can be as short as a float of a few inches, or as long as a hazardous drift of 20 to 30 feet or more.

After emergence, duns that escape feeding fish fly to nearby vegetation, where they await the transformation to the final and fertile spinner stage. Within a day or two they mate in the air, then the females deposit their eggs back into the river. I have not experienced an important spinner fall, though I know there must be times when trout take them.

March Brown

The dun is the most important stage of the March Brown, with exceptions that I'll cover later. The duns are fairly large insects; their imitations are typically tied on size 12 and 14 hooks, though sometimes late-season hatches are as small as size 16. They have two tails, tan to dark brown in color. The body of the insect is a dark reddish-brown, but its underside is much lighter, usually tan, and has a distinct olive cast. This does not show unless you tip the insect upside down and look at its underside, the way a trout sees it. The wing of the dun is mottled with brown patches. The head, if looked at closely, will reveal a flatness that is carried over from the nymph stage.

Mayfly duns lack distinct keying features. Professional taxonomists rely on male spinners for exact identification. Few anglers will want to acquire the taxonomic knowledge necessary to dissect and determine the species of a spinner. It is often best to identify a hatch by associating the emerging dun with the nymph from which it sprang. The easiest way to confirm a March Brown hatch is to collect a nymph. It will have three tails. Its head will be wide and flat, with eyes set into the top of it rather than out to the sides. Perhaps the key feature of the nymph is its overlapping gillplates, which form a suction-like disc on the underside of the abdomen.

Fly pattern rationale for the March Brown dun is linked to nymphal migration out of fast water. If it hatched where it lived you would need a heavily hackled dressing to imitate it. But it emerges into the quieter water near rapids and riffles, so flotation is not the problem it could be.

A dressing without a hackle floats well enough to take trout most of the time. But, as I said in the first chapter, I like a dressing that will fish on all the water types where I will find the insect hatching. That is why Bob Borden's March Brown Parachute fishes better for me than the more exactly imitative March Brown Comparadun. I like to be able to fish them in the rumpled water at the foot of a riffle or the head of a run.

The dressing for my favorite March Brown pattern follows.

MARCH BROWN PARACHUTE
Hook: Mustad 94840 or Partridge L3A, size

12-16.
Thread: Olive.
Wing: Natural brown deer hair, as parachute post.
Tail: Brown hackle fibers, split.
Body: Hareline Dubbin' No. 32, Olive Tan.
Hackle: Brown, parachute.

Since this dressing is Bob Borden's, it seems reasonable to call for his dubbin' in the tying of it. I have found that it works best, anyway, though you could substitute another olive-tan fur or synthetic.

As an alternative for smooth water, the March Brown Comparadun has split tails of brown hackle fibers or white nutria guard hairs, the same olive-tan body on olive thread, and a wing of tan or brown deer hair flared in a 160-degree arc over the body. It is an easier fly to tie, and you might well prefer it.

Tackle for fishing March Brown hatches does not call for anything out of the ordinary. I use an 8-foot rod for a no. 5 double-taper dry line. It's the same set-up I use for about 90 percent of my fishing, all season long. Your standard dry fly outfit will do as well for you. A standard leader 10 to 12 feet long, tapered to 4X, or 2- to 3-pound test, should be long enough to fool fish and stout enough to catch them.

Presentation is usually in water that is disturbed enough to allow upstream casts. I prefer to wade in from the side of a riffle or run, then fish up and across stream so that my fly goes over rising fish, but the leader does not. When fishing slicks and tailouts above riffles, it is often helpful to wade in above the fish, casting downstream to them with lots of slack in the line, so the fly comes to them ahead of the leader and line.

Earlier in this chapter I set March Brown nymphs aside, mentioning that they were important on occasion, and I left Rick Hafele standing in the water, just downstream from a long bright riffle. It's time to bring the two back, and get them together. It was because of Rick's knowledge of nymphal behavior that he outfished Tony and me so soundly on that Oregon river.

The March Brown nymph is important just before the dun emerges. But as old nymphs become new duns, more nymphs arise from the bottom. Most of us prefer to switch to a dry dressing when trout begin to feed on top. But trout continue to take a heavy toll of nymphs. The angler who fishes a nymph pattern properly can do great damage.

The proper nymph pattern need not be an exact imitation. The natural is a dark reddish-brown, usually a size 12 or 14. A Pheasant Tail Nymph in those sizes works well. A standard Gold Ribbed Hare's Ear will certainly take some trout. My favorite dressing for the nymph is a March Brown Spider soft hackle, tied with a hare's mask body, gold rib, brown partridge hackle, and orange tying silk. I'm not sure if it imitates the rising nymph or a drowned dun, but it works fairly well during a March Brown hatch.

But it doesn't work as well as what Rick Hafele uses. Rick's favorite dressing is a March Brown Flymph, and his dressing for it follows.

MARCH BROWN FLYMPH
Hook: Mustad 94840 or Partridge G3A, size 12 -14.
Thread: Crimson.
Tails: 2-3 pheasant center tail fibers.
Body: Hare's ear fur spun on crimson silk.
Hackle: Brown hen.

The proper way to fish this wet fly is within a few inches of the surface, with a cast that is across stream or slightly upstream. The drift of the fly should be closely tended with mends and tosses of line so the fly drifts downstream with a minimum of pull from the line and the leader. The fly should appear to be rising toward the surface and swimming feebly. Rick Hafele's ability to master the drift of his nymph, making it appear to be a rising nymph, was the key factor to his success.

The tackle required is exactly thta used for fishing the dry fly. A double-taper dry line is a distinct advantage, as it allows easier mending, and better control over the drift of the fly.

Rhithrogena hageni is a Rocky Mountain species of the March Brown that is so similar to *R. morrisoni* that entomologists must dissect out the male genitalia and count the spines on the penes before they can be sure of the difference between the two. Very few trout own the equipment needed to tell the species apart, and most will willingly accept the same patterns when either hatch is on. Their habitat overlaps; they often inhabit the same river.

I have fished hatches of *R. hageni* in June on Henry's Fork of the Snake River, just above the famous Railroad Ranch section. Flies tied for the Oregon *R. morrisoni* hatch that I had fished a few months earlier fooled the Idaho fish just as thoroughly.

Other Western *Rhithrogena* species include *R. doddsi, R. virulus, R. robustus,* and *R. undulata.* Most can be fished with the same patterns and tactics outlined here, though some will require slight variations in the size of the hook, and the color of the body fur used on the fly. But *Rhithrogena morrisoni* and *hageni* are the two species that comprise what is most commonly referred to as the Western March Brown hatch.

Little Olives

Baetis tricaudatus, bicaudatus, parvus, and Others

_____ Chapter 3

I t was April. I was on Oregon's Willamette
River with noted fishing writer Deke Meyer.
We had parked the boat on a gravel bar alongside
a long and slightly choppy run. It was a cold day,
with high clouds. There was no hail and no rain,
but we wore slickers to keep the wind from knifing
under our vests.

March Browns started emerging just after
lunch. Deke bolted off upstream a hundred yards
or so; I waded in at the boat and let the current
slowly urge me downstream. Soon the air was
filled with fluttering duns, and fish were taking
them off the surface with rises that sent jets of
spray into the air.

As I worked downstream I dropped my fly
along the edge of the main current, fished it down
a few feet on slack-line casts, then lifted it up to
place it on the water again. I did fine, taking an
occasional trout up to 15 inches long. I looked
upstream and saw that Deke was doing fine, too.
He usually does.

But suddenly the fish quit on me, though they
didn't quit rising. They just refused my fly.

Nothing had changed. March Brown duns still

boated the currents, still left the water to be scattered by the wind. I fished another half hour, but couldn't buy another rise. I looked upstream again. Deke was still doing fine.

I examined the water closely, and small bits of information began to fall into place. First, I watched a few March Brown duns emerge and float until they escaped into the air. None were taken, though fish boiled all around them. Then I thought of nymphs, but fish taking nymphs seldom toss spray into the air, and these fish still did that, though not quite so violently now. Finally I waded out as far as I could, until I was right among the rising trout. Then it was easy to see what they were taking.

A tiny olive dun popped up, got tipped to the surface by the wind, got taken by a greedy trout. Another came up, flew three feet, got blown to the water and instantly disappeared in a swirl. The closer I looked the more of the small olive mayflies I saw. The more of them I saw, the more I understood why the trout suddenly refused the March Brown dressing. They preferred the smaller insects, and concentrated on them ruthlessly.

I frisked my vest and located a tiny plastic fly box. One compartment contained a dozen perfect traditional dressings, given to me as a gift from Peter Rayment of Salem, Oregon's Santiam Flycasters. His flies were on size 18 hooks. They had tails of dun hackle, bodies of olive fur, wings of blue hen-hackle tips, and the kind of brisk medium-dun hackle that is damned hard to find.

I lengthened my leader with a couple of feet of 6X tippet, then tied on one of Peter's flies.

The change in my luck was almost instant. I worked back upstream toward the boat, casting to the quiet water along the edge of the run. Every time I succeeded in covering a rise I succeeded in hooking a fish. The rest of the hatch was a brief half hour, and seemed to be extinguished by the disappearance of a cloud that had obscured the sun. The bolt of bright sunshine turned the insects off. But I took five more fish in that short time.

When Deke and I got together at the boat I asked him what he had been using. His luck had been more constant than mine.

"March Brown dun," he answered. "I never switched."

"See any *Baetis?*" I asked.

"Not a one," he answered. Just one hundred yards away, in water that was not much different, the river provided exactly the hatch it was supposed to provide: March Browns. But down where I fished, the larger insects merely masked the hatch of Little Olives, which were preferred by the fish.

Little Olives hatch from March all the way through to November. There are two peaks, the first from April to the end of July, the second in September and early October. Early and late in the season, they emerge during the warmest part of the day, generally from 11 o'clock in the morning until about 4 o'clock in the afternoon. When the weather is warm, they tend to hatch in the cooler hours, usually at dusk or just before it. The best hatches are on cloudy days, when the insects seem to come off consistently for several hours.

Nymphs of Little Olives live only in moving water, but they have adapted to nearly all types of flowing water. There are good populations in some sections of all freestone streams. Sometimes they are isolated, as I found the day I fished the Willamette River with Deke Meyer. They are more abundant in meadow streams, especially where the currents are gently riffled in places.

The greatest populations of Little Olives are found in alkaline spring creeks. Densities in weedbeds can be so high that trout make a living

27

merely by rooting around with their noses and chasing the nymphs out into the open. A handful of spring creek weeds plucked out of the current can come up just wiggling with them.

Hatches of these small mayflies occur on virtually all Western streams. There are several closely related species; some live in the slower and warmer waters along the southern tier of Western states, others in the colder waters of the Sierra, Cascade, and Rocky Mountain states. If there is a water type where they are most important, it is on the many spring creeks with rooted weedbeds and smooth currents.

The life history of the *Baetis* mayflies is somewhat unique. They have two or three generations per year. One brood comes off early, and deposits its eggs into the water. The eggs hatch, the nymphs forage on algae and vegetation for a few weeks or a few months. Then they emerge again. If the water is warm enough and the season long enough, another generation may follow before the final brood that overwinters until the next spring.

This multi-generation life cycle is the reason more than one peak per season happens with Little Olive hatches. It is also the reason the peak of the hatch is hard to predict on a blanket basis for waters throughout the West: each river has its own temperature regimen, and each has its own hatch timing. You can get surprised by Little Olives nearly anytime, throughout the entire season.

The nymphs emerge by swimming to the surface in open water. Because they are so small they sometimes have trouble penetrating the surface film, especially on smooth currents with correspondingly high surface tension. Once emergence is complete the duns fly to the streambank, where they remain in the vegetation until they smolt into the spinner stage and return to mate and lay their eggs.

Spinner flights usually take place in late afternoon or evening, though they often occur in the morning during warm weather. The females of some *Baetis* crawl under water to deposit their eggs on submerged rocks, logs, or plant stems. This is rather adventurous behavior, and suggests the use of a very small traditional wet fly to imitate them. But it has been my experience that Little Olive spinners are not often important to the fly fisherman.

The dun is the most important stage. But the nymph, at the moment of emergence when it reaches the surface and tries to penetrate the film, is also extremely vulnerable to trout. Imitations of the insect at this disastrous instant can be more effective than dun dressings on quiet spring creeks.

Little Olive Dun

Little Olive duns are two-tailed. They have olive to tannish- or brownish-olive bodies. Their wings are blueish-olive or light gray; another common name for them is Blue-wing Olives, which is perfectly valid except that a lot of other mayflies creep in with the same name. Typical size range is from size 16 down to size 20. But they can be as large as size 14, especially in the first brood in early spring, and as small as size 24, especially in the last brood in late fall.

Fly pattern rationale for the Little Olives is based on their adaptation to such a wide array of Western water types. I have used Peter Rayment's traditional Little Olive tie with great success on fast water. I have used tiny comparaduns, the non-hackled ties, on spring creeks with happy results. But because the naturals are so small, I have constantly longed for a pattern that would float low in the water but show up well wherever I fished it.

One day a couple of Aprils ago, a few friends and I fished Armstrong Spring Creek in Montana's Paradise Valley. The fish were alarmingly selective, feeding on a sporadic hatch of *Baetis* that went on all afternoon. I'd had a bad winter for tying; my fly boxes were empty of anything that would work for the tiny insects. It was a frustrating afternoon.

The next day we had rods on Nelson's Spring Creek, just across the Yellowstone River valley from Armstrong. We spent the night in separate rooms in a cheap railroad hotel in Livingston. I set my alarm for 6 o'clock in the morning. When it went off I struggled out of bed, clamped a tying vise to a night stand, and sat tying size 18 and 20 flies by weak lamplight.

About 7 o'clock I looked out the hotel window and saw Rick Hafele, the well-known angling entomologist, walking down the street toward the bus station cafe for his morning coffee. I had only half a dozen flies tied.

An hour later Jim Schollmeyer, the photographer whose beautiful black and white pictures show up in so many angling magazines today, walked by the window to join Rick for coffee. I had a dozen flies tied.

At 9 o'clock Richard Bunse, illustrator of angling books, strolled beneath the window in the same direction, and I knew it was time for me to finish the last fly and catch up with my friends for breakfast. The food in the greasy spoon was plentiful and good. And I had 18 flies to try over the *Baetis* that day. All were parachute ties, and all had white topknots that I hoped I could see from 40 to 50 feet.

On the creek, the hatch started just after noon. I cast over lots of rising fish. I didn't catch all of them, or even half of them. But I caught my share, and when I did my part right the fish rose and took the fly the way they were supposed to.

My favorite fish rose in an indentation among the willows, on the far side of the creek in a wide place where it couldn't be waded. It was 50 feet from where I stood to where the fish came up with a steady rhythm. The slot in which it worked was only about three feet long, guarded by willow limbs above and below. I would have to drop the fly at the upper end of the slot just as the trout was ready to rise. Then I would have to get a perfect float while the trout made up its mind.

It was the kind of cast that better fishing writers can make every time, and probably wouldn't even stoop to writing about. But I haven't got my prose worked out that well yet. I made the cast from slightly upstream. The fly surprised me by landing gently, just inches below the upper willow branch. It drifted primly to the fish, and was taken in a

31

satisfying swirl. It was only a two-pound rainbow, not much of a fish for the creek it came from. But the confidence of its take justified the three sleepy hours I'd spent at the tying vise that morning.

I spent the rest of the afternoon losing the rest of the flies to willow limbs and to hooked trout diving into weedbeds.

Here is my dressing for the Little Olive Parachute.

LITTLE OLIVE PARACHUTE
Hook: Mustad 94833 or Partridge L3A, size 16-22.
Thread: Olive.
Wing: White calftail, tied as post.
Tails: Dun hackle fibers, split.
Body: Olive-dun fur, Hareline No. 29
Hackle: Blue dun, tied parachute.

I like the light-colored olive-dun for the body because it is the closest I have found to the spring creek hatches I have encountered. I like to match those as closely as I can because trout in that type of water are more selective than trout in more tumbled water. Alternatives to the olive-dun are an olive-brown fur (Hareline no. 30) or an olive and hare's ear mixture (Hareline no. 34).

For those who fish only on spring creeks, the comparadun dressing might be a better choice. It has split tails of dun hackle fibers, an olive, tannish-olive, or grayish-olive body, and a natural dun deer hair wing flared over the body. This gives a more exact silhouette of the natural on smooth currents, but is more difficult to see on water that is at all rough. I prefer a fly that covers all the water where I might encounter Little Olive hatches.

Tackle for fishing these small flies should be fine. A no. 4- or 5-weight line works well. Leaders should be tapered to 5X or 6X, one to two pounds test, and should be at least 12 feet long for smooth water.

Presentation can be upstream, or up-and-across, on riffles and runs. But on spring creeks it is far better to work from a 45-degree angle above the fish, casting with lots of slack in the line, so the fly drifts freely down the feeding lane ahead of the line and leader.

Matching the emerging nymph can be just as effective as matching the dun where the water is smooth. The best dressing floats in the surface film, like a trapped natural, and embodies a knot of fur or polypro yarn to copy the partly emerged wing of the dun.

The nymphs are slightly darker than the duns, usually a brownish-olive. Some species have two tails; others have three tails, with the center tail shorter than the other two. There are tiny gill plates along seven of the abdominal segments. The head is hypognathous, which is a way of saying that it points down instead of straight ahead. This posture helps the nymph browse on aquatic vegetation.

A dressing for a killing emerger pattern follows.

LITTLE OLIVE FLOATING NYMPH
Hook: Mustad 94833 or Partridge L3A, size 16-20.
Thread: Olive.
Tails: Blue dun hackle fibers.
Body: Brownish-olive fur.
Wing clump: Knot of brownish-olive fur or gray polypro yarn.
Legs: Blue dun hackle fibers.

The wing clump should be dressed with floatant to keep the fly in the film. The fly should be fished exactly as if it were a dry. Drift it with a drag-free float over rising trout. Hit them right on the nose if you can; they will not move far to take it when lots of naturals are on the water.

In the kind of glassy water where this dressing is effective, you will almost always want to present the fly from above the fish, rather than below it. Success calls for long leaders, delicate casts, and the kind of stooped posture that keeps you out of sight from the fish, in agony with a sore back when the fishing is over.

Hatches that are similar to the Little Olives of the *Baetis* genus include the sister genus *Pseudocloeon*. These are generally a smooth-water hatch, though they are found in calm currents on free-stone rivers. They are difficult to distinguish from *Baetis,* though they tend to be slightly lighter in color, and slightly smaller in size. The Little Olive Parachute dressing given should fish well for them when tied with the olive-dun fur on size 20 through 24 hooks.

They emerge afternoons, June through September, and are abundant on many famous Western waters, including spring creeks like Henry's Fork of the Snake River. They are so similar to the Little Olives that you will likely never know the difference, and will not find it necessary to adjust either your tackle or your tactics when fishing over them.

Speckle-Wing Quill

Callibaetis nigritus, coloradensis, pacificus and Others

Chapter 4 _____

I t was the early days of the float tube movement, the dying days of the Hippie movement.
There was an untidy encampment on my favorite
lake when I arrived. It looked like they'd suffered
the winter there. I avoided the camp, inserted
myself quietly into my waders and fins, then into
my tube, then into the water. I paddled away
around a point where I thought they couldn't see
me.

It was late May. The sun, which now and then
hid behind a cloud, felt good whenever it was out.
Nothing was happening on the lake, so I dropped
a weighted nymph over the stern and towed it idly
around for a while. Nothing touched it.

About 11 o'clock a few rises broke to the surface. But there was no apparent cause for them, so
I didn't switch from my unsuccessful tactic. After
a few minutes some duns began to pop out on the
surface. A couple disappeared in the kind of rises
that leave ridges on the water. I switched to a dry
fly.

I'd fished this hatch often, and needed no close
look at the dun to know how to match it. While
the rises were still sporadic I caught my first fish.

Soon the water was disturbed all around me, and my fly flicked out this way and that to cover the closest trout. There were few refusals. From shore I'm sure that I appeared to be an expert fisherman.

I was under observation from shore. Two grubby looking characters crept out of the trees and hailed me from 75 yards away. I turned to look at them while I played a fish. They carried spinning rods with bobbers almost the size of Chlorox jugs dangling from the tips. I wondered what kind of ammunition they were tossing at the trout.

"What're you catching out there?" one of them called.

"Fish!" I shouted back. I was angry at being disturbed, and these were bait fishermen.

"What kind of fish?" he asked, not angry.

"Trout," I said.

"What're you catching them on?"

"Flies," I answered. That much should have been obvious, but apparently the movements of fly fishing were foreign to him.

"Oh," he said, stumped. He didn't know what questions to ask a fly fisherman. Finally he said, "What color flies?"

I answered, "Brown." It didn't mean much, and they faded back into the trees.

I was relieved when they were gone. But they had been polite enough, and I hadn't. I should have been ashamed. Their pants were nearly falling off; they looked like they needed a meal, and I'd released 15 trout while they watched. But I don't know how they would have fished flies with Chlorox jugs, if I'd offered them any.

Speckle-wing Quills start in the low-lying coastal areas as early as the last week of April. The hatch strengthens through late May and early June. In some high lakes in the Rockies it doesn't start until July, but the ice doesn't melt in some lakes until then, and insects have trouble emerging through the surface film before ice-out. In lakes where the hatch starts early, it continues through the season, into October, with peaks in late spring, early summer, and again in fall.

Emergence starts daily a little before noon, though early in the year when the water is still cold it might start an hour or two later. On bright days the hatch period is brief, sometimes just an hour or so. But when the sun is hidden by clouds the hatch can go on for three to four hours.

Speckle-wings are closely related to the Little Olives covered in the last chapter. Unlike the smaller *Baetis* genus, which are restricted to flowing water, *Callibaetis* inhabit lakes, ponds, and only those streams with very slow currents. Their populations are greatest in still waters with dense weedbeds.

They are distributed throughout the West. Though they inhabit some spring creeks, their primary importance is in lakes and ponds. The early peak of the hatch often occurs when streams are out of shape due to spring runoff, making them a hatch to turn to at a time when river fishing is poor. Strangely enough, one of the best hatches I have ever witnessed was in the flooded rice fields of the Central Valley in Califoria. It wasn't much of a trout fishing opportunity, though, so I shall not report on it here.

Like the Little Olives, Speckle-wings have two and sometimes three generations per year. That is why the multiple peaks in hatch intensity. In general, if there is a hatch in May and June, there will be another in late July, and a final appearance in September. But these regimens must be worked out for every lake, and cannot be pinned down for the whole West.

Speckle-wing nymphs are strong swimmers,

darting about in the vegetation and perching cockily on it. They feed on the thin layer of bacteria and algae that grows on plants and debris. When mature they swim boldly to the surface, penetrate the film quickly, and emerge into the dun stage. If the weather is warm they escape the water quickly. If wind or rain make flying difficult, trout get a longer shot at the duns.

After the molt to the spinner stage in lakeside vegetation, Speckle-wings return to the water to mate above it and deposit their eggs in it. The clouds of dancing males can be so dense that specimens soon cover your shirt and hat. They generally start around 2 o'clock in the afternoon, just after the emergence, and go on into early evening.

The nymphs are available as they ascend to the surface. Stomach samples over the years have confirmed that trout take more nymphs than duns. But I prefer to fish dry, and consider the dun the most important stage of this important lake-dwelling insect.

I have read and heard reports that the spinners are most important. But it has been my experience that the spinners are largely ignored by trout, at least where I've fished these hatches. Such statements will get me in trouble with those who have fished spinner patterns successfully, but it is best that I write what I've seen rather than what I've read or heard.

Speckle-wing Quill duns are fairly large. The first generation usually arrives at about a size 12 or 14. Each succeeding brood is a size or so smaller. Solving the hatch in May is not the same as solving the hatch n September. You'll have to use an imitation two hook sizes smaller, a size 16 or even 18.

Duns of this group have two tails. The body is tan, brown, or grayish on the back. But as usual among the mayflies, the underside will be lighter, and will have an olive tone added to the dominant color, or may even be the dominant color. The wings are tan to brown, with a distinct mottling that is unmistakable.

Speckle-Wing Quill Dun

Fly pattern selection for the Speckle-wing Quills is not as difficult as it has been for the groups already discussed. These still-water types eliminate the need for great flotation; it is best to concentrate on better imitation of the silhouette of the natural.

I was not being entirely truthful when I told those hungry Hippies that I fished with a brown fly. The naturals I imitated were brown on the back, but on their undersides they were tannish-olive, which is true of the common run of Speckle-wings I have collected. The best dressing I have found for them is the *Callibaetis* Comparadun. The dressing for it follows.

CALLIBAETIS COMPARADUN
Hook: Mustad 94833 or Partridge L3A, size 12-18.
Thread: Olive.
Wing: Mottled brown deer hair, in 160-degree arc over body.
Tails: Ginger hackle fibers, split.
Body: Olive and tan fur, mixed.

Though I have taken fish on Light Cahills, and even Ginger Variants, during hatches of Speckle-wing Quill duns, it is far better to cast the more accurate comparadun when fish are at all selective. It floats flush in the surface film and gives a more realistic impression on the water.

It is well to remember that you must use a smaller dressing as the season progresses. The same tie in a range of sizes will work on most lakes and ponds throughout the West. In fact, the same fly will work very well for the earlier March Browns, wherever they hatch on water that does not require hackle to float the fly.

Tackle for fishing these hatches should take into consideration the often heavy winds on Western lakes. I like a long rod for a no. 6 line, which helps me put out long casts from the low position of a float tube. Leaders should be 12 to 15 feet long, tapered to 4X or 5X, 3- to 4-pound test. If your lakes are like my lakes, the chance of an occasional large fish does exist, and I like a leader strong enough to give me a chance against them.

When presenting the flies the only thing to consider is getting the fly out to where rising fish can see it. Let it sit there a bit. Let the fish come to it. If you try to paste it into every rise ring in sight, you'll soon have all the fish frightened around you. The biggest problem I have is dozing off while the fly sits during a sporadic hatch. When a fish hits I'm not ready, and I set the hook long after the fish has gone wandering on.

Many Speckle-wing nymphs are taken by trout as they make their way to the surface to emerge. Imitation of this stage can be very effective, though I prefer to fish dry if trout will take on top.

The nymphs are generally shades of tan or brown with some olive thrown in. They are slight chameleons, and will take on some of the shades of the surroundings in which they live. They have three tails of equal length. The gills along their abdominal segments are heart-shaped, and some gills have smaller recurved flaps under them that look like second gills. The antennae are long, and that separates them from the only other mayfly nymph you might confuse them with in still water, the *Siphlonurus*, which have short antennae.

A size 12 to 16 Gold Ribbed Hare's Ear will take some trout during a *Callibaetis* hatch. But the best nymph pattern I have found is one that shouldn't work. It is a Hare's Ear wet fly. The dressing for it follows.

HARE'S EAR WET
Hook: Mustad 9671 or Partridge H1A, size 12-16.
Thread: Primrose yellow silk.
Tails: Wood duck flank fibers.
Rib: Narrow gold tinsel.
Body: Hare's ear dubbing, or Hareline No. 4
Wing: Partridge or hen pheasant wing quill
    segments.

This is a traditional winged wet fly dressing that has been around longer than I have, and will be

42

around longer than I will. The rough fur dubbing, with the guard hairs left in, serves as hackle.

The reason it shouldn't work is the wing. Rising Speckle-wing nymphs don't have their wings trailing. But I've used this fly during their hatches, and it works. Catching fish is one of the primary elements of fishing; when I find a fly that does that, I like to know why, but don't require it. I'd rather use a fly that shouldn't work but does than a fly that should work but doesn't. I'm not very sophisticated.

Presentation of this dressing is just as simple as the reasons I use it. I just cast it out, let it sink a bit, then retrieve it back with short strips. It must look like a nymph swimming along, looking upward, trying to find the right hole through which to penetrate the surface film. I use the same line and leader I use when fishing dry during the same hatch.

Western Green Drake

*Ephemerella grandis,
glacialis, and doddsi)*

Chapter 5 _____

It was a humbling lesson in selectivity. Richard
Bunse and I fished Oregon's lower Metolius
River. The upper river is a classic bemeadowed
spring creek, meandering gracefully through
stands of great ponderosa pines. The lower river is
a different critter. It plunge whitely down through
mile after mile of angry rapids. There are perhaps
a dozen places where one can timidly wade the
edges of it. Fish should not be fussy in water that
mean.

We fished it in mid-June. The Green Drake
hatch was nearing its end. Only a few duns were
left on the river; only a few fish rose to take them.
Bunse, who knows the water better than I do, put
me into a place where it was possible to wade at
the side of a swift run and cast out over its holding
water. He waded in 50 yards below me, and we
both began fishing.

I used a heavily hackled Wulff dressing with a
green body and dun hackle. It has worked well for
me when Green Drakes come off in coastal
plunge-pool streams. But it didn't work on the
lower Metolius. I cast it for two hours and didn't
get a single take.

Bunse, fishing below me in water that was exactly the same, didn't have fishing that was a lot faster than mine. He used a new fly of his own design, an extended tie with a body made from buoyant packing foam. It looked a lot more like the natural dun, and it bobbed on the water like a tiny inflated raft. Four trout rose to take it. He was able to hold only one of them against the brutal current. It was a 3-pound rainbow, nearly 20 inches long. The ones he lost were larger.

A week earlier Bunse had hooked a fish in the same place. When it turned its side to the fast water and started down with the power of the river behind it, Bunse's rod snapped just above the handle. That's the kind of fish that get active when the Green Drakes are on, in lots of Western rivers.

Green Drakes hatch in late May and early June in the West coast states. In the Rocky Mountain region, where the hatches are perhaps most famous, the hatch starts around the end of June and continues on various waters through the end of July. It is loosely keyed to a 50-degree water temperature, but too many other factors work in the equation to make it exact.

The hatch might last three to four weeks on some waters. But in most places it takes a few days to warm up, then peaks for only two or three days before tapering off again for a week or so. A lot of wadered anglers sit sadly alongside famous rivers every year, their rods still unstrung, disappointed because they've wasted vacations trying to hit the peak of the Western Green Drake hatch.

But if you go fishing whenever you can, and let the hatch take care of itself, you are going to run into it on one Western water or another. It hatches on nearly all of them.

Daily emergence usually starts around 11 o'clock in the morning. On sunny days it might last only until 1 o'clock. When clouds cover the sun the hatch can go on until as late as 4 o'clock. If the water is warm the hatch might begin late in the afternoon and trail on into the evening.

Green Drakes are distributed throughout the entire West, from New Mexico to Alberta, from California to Alaska. They are dominant, and perhaps the season's most important hatch, in most cold-water streams. The most famous hatches are in our mountain states: Colorado, Wyoming, Montana, and Idaho.

Green Drake nymphs live only in moving water. They have adapted to almost all kinds of streams, but their best populations occur either in well-oxygenated fast water, or in spring creeks that have constant cold flows. Henry's Fork of the Snake, in Idaho, is home to the singlemost famous hatch. I have fished over great numbers of a closely related species, *Ephemerella coloradensis,* on the frigid waters of Fall River, an Oregon stream that emerges from the ground at 42 degrees, year around.

The life cycle of these large mayflies lasts one year. They quit the egg soon after it is dropped to the water in early summer. They forage as nymphs among bottom stones or weedbeds for the rest of summer and fall. They overwinter and continue to grow through the cold spring months. As the water approaches 50 degrees they prepare to emerge.

Emergence takes place either in pockets in fast water, or else out in the fast water itself. The nymphs are very awkward. Their trip to the surface is half swim, half drift. Most shed the nymphal shuck before they reach the top, and are then at the mercy of the current until they are buoyed up to the film. Though this type of emergence suggests the use of a wet fly or flymph, I have not experimented in that direction yet, and cannot offer, from my own experience, any useful dressing to imitate the nymph or emerging dun.

Green Drake duns usually suffer an extended drift on the surface while they wait for their wings to dry. They flutter and try to take off, fail, drift a ways, then flutter until they finally escape the water. Trout like this. They get greedy when it happens, and gorge themselves on the helpless duns.

After a day or two in streamside vegetation the molt to the spinner stage is complete, and the fertile adults return to mate in the air and lay their eggs. This is reported to take place at night. I have not encountered any fishable spinner falls of the major Green Drake species.

The most important stage of the insect is the dun. It is extremely vulnerable to trout as it flutters along. It is so large that it draws up the largest trout in the stream. Its great disadvantage is that it is difficult to imitate a dun so large. Trout tilt the most selective noses of the season when the Green Drakes are on.

Green Drake Dun

47

The duns are sometimes a size 10, though a size 12 is more common, and a size 14 not rare. They have three tails and portly bodies. They are dark brownish-olive on the back, but the underside, which we must imitate, is almost a bright olive. There are distinct yellowish bands between the segments. The wings are a dark slate gray. The hindwing of the Green Drake displays a projection along its leading edge that aids in identification of the natural.

It is important to remember, with this hatch, that the dun will darken soon after it leaves the water. If you are to see what trout see, you must collect a dun just at emergence and observe it immediately. The brightness of its green will begin to fade to brown within an hour or so.

Fly pattern rationale for Green Drakes has emerged more from defeat than from victory. I am not often an advocate of exact imitation, but this is one hatch where approximation has not served me well. It is the patient Bunses of the fly tying world who have the most success over these large insects. I'll get to his fly later.

Mike Lawson owns Henry's Fork Anglers, a fly shop on the banks of the river that contains the most famous Green Drakes, and the most famous selective trout. He created the Green Paradrake to match the hatch on his waters. It is the accepted dressing wherever anglers gather to fish this difficult hatch. The dressing for his fly follows.

GREEN PARADRAKE
Hook: Mustad 94840 or Partridge L3A, size 12-14.
Thread: Yellow.
Wings: Natural gray deer hair, as post.
Tails: Dark moose hairs.
Body: Dark olive-dyed elk hair, ribbed with working thread.
Hackle: Dark dun, tied parachute.

The Lawson tie is effective over the most selective trout. Since so many people choose to seek out this hatch on storied waters, it is best to be prepared to fish it with the best flies you can tie or buy.

A comparadun tie can be effective where the fish are slightly less selective. It is tied on a size 12 or 14 fine-shank hook. It has split tails of beaver or nutria guard hairs. Its body is a mixture of brown, olive, and yellow dubbing furs. The wing is a natural slate-gray deer hair, flared in a 160-degree arc over the body. It is distinctly easier to tie than the Paradrake.

The fly that Richard Bunse used on the Metolius is new, but it is no longer experimental. He ties them all winter, but his supply in spring fails to dent the demand at the Camp Sherman store, on the river. Because it is so new, so difficult, but so effective, I will give the tying steps for it at the end of this chapter. If you are a patient tyer you should try a few. You won't be disappointed. It was originally called the Wonder Dun, but an embarrassed Bunse has asked that its name be toned down to the Natural Dun.

It takes relatively stout tackle to propel such large flies. Use at least a size 5 line; I prefer a 6, especially for the winds that often attack the rivers during this hatch. Leaders 12 to 15 feet long are helpful on spring creeks, though shorter leaders are fine where the drakes come off on rougher water. A tippet of about 4X is just right, turning over the flies while being fine enough to fool the fish.

Upstream casts will work on tossed water like the lower Metolius. But you won't catch many fish without working downstream to them on such glassy waters as the Henry's Fork. Work into a position at an angle above the fish. Cast right to the feeding lane, a foot or two above the rise. Waggling the rod while the line is in the air causes it to land with a series of S curves. As these straighten out the fly drifts naturally over the fish.

There is a group of slightly smaller species closely related to the Green Drakes. They are variously called the Little Green Drakes and the Blue-wing Olives. They include *Ephemerella coloradensis, flavilinea,* and *spinifera.* The listed dressings tied on size 14 and 16 hooks fish well for them. I usually use comparaduns with split guard hair tails, olive bodies, and natural dun deer hair wings. It is wise to have these handy in a corner of a fly box, as it will be a poor season when you don't encounter some sort of small Blue-winged Olives on the water, wherever you go in the West.

The dressing for Richard Bunse's Natural Dun follows. The photos and tying instructions are courtesy of Deke Meyer. They originally appeared in the October, 1986 issue of *Flyfishing* magazine, in an article written about the fly.

NATURAL DUN
Hook: Mustad 94838 or Partridge L3A, size 12.
Thread: Yellow.
Wing: Natural dun deer hair.
Tails: Nutria or beaver guard hair fibers.
Body: Packing foam from a stereo or TV, colored
   with magic marker and sprayed with waterproof
   artist's fixative.

Tying Instructions:
1. Insert sewing needle in vise. Cut foam in an elongated shape. Cut a groove in the foam lengthways for the needle. Color foam with magic marker, spray with artist's waterproof fixative.

2. Wrap thread on needle about halfway. Place foam on forward half of needle, catch end of foam with thread and tie down snugly, but not too tightly to eventually remove needle. Tie in tails at tie-down area, but don't trim excess.

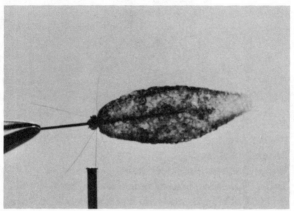

3. Pull foam back out of the way, tie down tails while bringing thread forward to first segment tie-down area. Pull foam forward, go around foam with thread one time and tighten. Continue with each segment: pull foam back out of the way, wrap thread up to where the next segment will be, bring foam forward, go around the foam one time. Half hitch and clip thread at the last segment.

4. Slide foam body off needle, remove needle from vise, insert hook. Attach thread and tie in stacked deer hair wing, comparadun style, trim, half hitch thread but leave attached. The hair wing should be one-and-a-half times hook length (on a short shank hook) and tied in at the middle of the hook shank.

5. Take hook out of vise, run hook through the foam body just in front of the last segment, in the groove, with the groove on top, so the hook comes out the bottom. Put hook back in vise, take three tight turns over last segment tie-down area.

6. Wrap thread forward, up to eye of hook. Trim tail butts if necessary. Hold deer hair up with fingers, while pulling foam body forward and up around hair wing at the same time, then wrap thread over foam body several times at the eye of the hook.

7. Trim off excess body material in front of fly, wrap down with thread. Cut slit in the foam body to the sides of the deer hair so the wing can fan out to 180 degrees, but don't pull wings into slit yet.

8. Bring thread back, wrap two times in front of wing, then behind wing two times to push body up against wing, but not collapse body too much. (Stay out of wing slits with the thread.) Bring thread forward to hook eye and whip finish. Pull wings into side slit, forming 180-degree fan. Cement base of wings, head, and entry point of tails.

Salmonflies

Pteronarcys californica

Chapter 6 ————————————

I was floating down the Deschutes River one
scorching June noon. I glanced toward shore
and saw a fellow standing bent over, alongside a
bunchgrass bank. Both his hands were in the
water. It was obvious he was reviving a fish and
about to release it.

"Lift it up," I called across to him, "so I can
see it."

He did. A month later I'd have thought it was a
summer steelhead. It was 20 inches long, one of
those chunky native rainbows that make a football
look as elongated as a link sausage. The fellow
lowered the fish back to the water, held it for a
few more seconds, then straightened as the fish
swam powerfully away.

I pulled ashore to fish the banks not far downstream. I used the normal run of dry salmon fly
dressings. None of them worked, though great big
naturals lumbered awkwardly through the bunchgrass clumps and all along the limbs of willows.
One crawled across my glasses. There were enough
around that the trout should have shown some interest. They didn't. I gave it up and sat on the raft
in the shade.

After a while the same fellow who had taken the big one came along.

"What did you take that trout on?" I asked him.

"A Langtry Special," he answered. "It imitates the salmonfly."

"Never heard of it," I said. "Can I have a look?"

"Sure," he answered. "Here, have one." He gave one to me. I took a chance and whined a little about how I'd be in serious trouble if I had only one and lost it. It worked; he gave me another. He introduced himself as Jim Meyers, and said that he was from Corvallis, Oregon.

Mr. Meyers told me the fly was originated by Judge Virgil Langtry, of Deschutes River fame, and that it had done serious damage to the trout population for years during the salmonfly hatch. I'd witnessed some of the damage, and was eager to try inflicting some of it myself.

I rafted downstream a polite distance, went ashore, and worked my way into an indent in the willow shoreline, where the current was deep and pugnacious along a rip-rap bank. It's the kind of water that holds large and feisty trout, protecting them by being almost impossible to cast to.

I crouched at the bank and dropped the fly on the water at my feet while I worked line off the reel and sized up the forecast and backcast areas. While I calculated my first cast, a trout came up and jumped the fly. It was about 14 inches long. I landed it and released it and made a note to myself that a fly has some promise when trout will take it at your toes.

A few casts later I worked up to where the fly landed right under the sweep of an overhanging willow, the kind of place where awkward salmonflies clumsily fall to the water, and trout eagerly await them. There was a boil at the first cast, but the fish missed the fly, or I missed the fish. Usually when this happens the trout won't come again. But this one did, on the next cast. One of us missed again.

On the third cast the frustrated trout murdered the fly. A few minutes later I bent over to hold it in both hands, to revive it before releasing it. It was nearly, but not quite, the size of the fish Jim Meyers had released earlier while I rafted past.

I fished the Langtry Special for the next two days of the float trip, and it took trout more consistently than any other salmonfly dressing I have used when those awkward beasts are about. I've been using it ever since, with similar luck.

Giant salmonflies begin to hatch when the water reaches a bit above 50 degrees. This can be as early as April on some California and Oregon rivers west of the Sierras and Cascades. In rivers east of these mountain ranges the hatch begins in late May and travels on for the first couple of weeks in June. It begins in the first week of June on rivers in Idaho, Utah, and Colorado. In most Montana and Wyoming rivers the salmonflies begin in late June and continue to hatch well into July.

Though it is the largest insect in most Western rivers, the salmonfly loses importance on many waters because its emergence occurs during spring runoff. Its greatest importance is in rivers that have protected flows: the Madison stabilized by Hebgen and Quake lakes, the Deschutes tamed by its dams, Henry's Fork of the Snake insulated from high water by Henry's Lake, Island Park Reservoir, and the spring creek nature of its major upstream tributaries. All of these have heavy salmonfly populations; all are clear and fishable when the hatch begins to happen.

Salmonflies are distributed throughout the entire West, living in most flowing water that is well-oxygenated. They are most abundant in medium to large rivers with lots of heavy riffles and rapids. They can be important wherever they hatch, so long as the rivers are clear enough that trout will move to take their imitations.

The life cycle of the salmonfly makes it important in both stages: the nymph and the adult. They live as nymphs for three and sometimes four years. The first- and second-year class nymphs are out there in the water all year round. Since they are larger than most other nymphs even in their youth, they are constantly taken by trout, no matter the time of year.

Once I fished the Deschutes with Michael Saracione long after the hatch was over. I had the third seat in the drift boat on a trip that Mike guided for Sylvester Nemes, author of *The Soft-Hackled Fly*. Since Nemes and his book are two of my heroes, I was honored to be on the trip.

Sylvester and I stuck determinedly with his soft-hackled wets all morning. But the Deschutes River trout denied us. When they won't come up, they won't come up; they are notoriously stubborn.

Mike is one of what I fondly call the "holy rollers." They are a Portland, Oregon crew who have mastered the technique of drifting a salmonfly nymph along the bottom of Deschutes River runs and riffles. Nobody is better at it, and there is no better way to take reluctant trout.

While Sylvester and I worked a pleasant riffle with no result, Mike snuck off downstream. When he was almost out of sight he limbered up his long rod and began rolling a salmonfly nymph along the bottom. Every time I looked down his way he was playing another trout. He took only four or five before he reeled up and came wading back upstream with a guilty look.

"I shouldn't be fishing these ugly nymphs with you guys around," he said. "Let's drift on downstream.'' It was the first time I'd heard anybody apologize for catching trout.

Salmonfly nymphs spend most of their life cycle clambering along the bottom in fast water. They are herbivorous, eating algae that grow on bottom stones, what rooted vegetation there might be down on the bottom in fast water, and the leaves and other detritus that gets trapped among the rocks. When it comes time for their emergence, they migrate toward shore in great numbers, sometimes great masses.

Salmonflies crawl out of the water before the adult casts the nymphal cloak. This almost always takes place at night. The moment of emergence itself is of little value to fishermen. But the migration just before it is a remarkably important event for the angler. Fish apparently consider the nymphs to be fine groceries; they seem to key on them until well into the hatch, refusing even a well-presented adult dressing so long as they can get a full feed of nymphs.

Salmonfly Nymph

Salmonfly nymphs are easy to recognize. They are large; the second year class will be about an inch long, the mature specimens nearly two inches long. They range from a dark reddish-brown to a brown that is almost black. They possess two short tails, and two long antennae. The tips of the wingpads are pointed. On the underside of the thorax, tufts of white gills look like graying chest hair.

There are any number of complicated dressings to imitate the salmonfly nymph. I created one myself once, and wrote an article about it—one of those writings that jumped to an early conclusion. But the truth is, you don't have to imitate every detail of the insect in order to interest the trout. The best flies are simple, and I haven't found one better than Mims Barker's Box Canyon Stone, not even my own. His dressing for it follows.

BOX CANYON STONE
Hook: Eagle Claw 1197B, size 2-8.
Thread: Black.
Weight: 12-15 turns of lead wire the diameter of the hook shank.
Tails: Dark brown goose quill fibers tied in a V.
Body: Black yarn, twisted to give segmented effect.
Wingcase: Brown mottled turkey section tied in over the thorax.
Legs: Furnace hackle wrapped over thorax.
Thorax: Black yarn.

It takes stout gear to cast these large, heavy flies. Nine foot and longer rods are good, as they keep the lethal things well away from your head. The line should be at least a no. 6, with most people preferring to fish heavier 7's or 8's.

If the water is shallow enough, use a dry line, an 8-ft. leader tapered to about 6- or 8-pound test, and a strike indicator a couple of feet above the fly. This gives you more control over the drift. But it will often be necessary to use a Wet Tip or even Wet Head line to get the fly down where you want it. In that case, use a leader about three or four feet long, and forget the indicator unless you want to dash your head under the water to watch it.

Casts should be short, and upstream, or upstream and across. Fish no more than 40 feet of line so you can control the drift. Let the fly tumble back down, drawing in line as it does. When it passes you, start feeding line out again to extend the drift as far as you can downstream. Sometimes takes will come right under the rod. They come most often just at the point in the drift when the fly lifts off the bottom.

Takes will be somewhat subtle most of the year. You'll be glad of your strike indicator. But during the nymphal migration, when trout seem to be racing each other to get at the tumbling naturals, the take will often be brutal. You'll never get a chance to set the hook. You won't need to do anything but dig your heels in and hold on to the rod.

Salmonfly nymphs crawl out to emerge at night. The adults stick around to rest and mate in streamside vegetation for several days. The result is a period of about three weeks to a month when they are available to trout.

Their greatest area of vulnerability is their awkwardness. They hang along willow limbs and grass stems right at the river's edge. They do fairly well as long as they hold still. But when the sun gets warm and their engines warm up, they get excited and go gadding about. Half the time they wind up falling into the water.

The best fishing for them is right along the banks throughout the afternoon, but in late afternoon or evening mating flights take place. These are sometimes sporadic, and draw little interest from the fish. When lots of these big insects hit the riffles, however, not many fish ignore them.

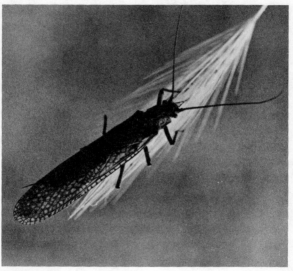

Salmonfly

There are not many things that you can mistake for a salmonfly adult. They are a couple of inches long. They have two short tails, two long antennae. Their wings are relatively flat, and heavily veined. The body is dark on the back, but the underside is orange. There is usually a fiery-orange band behind the pronotum, which is the first segment behind the head.

Here is the dressing for the Langtry Special.

LANGTRY SPECIAL
Hook: Mustad 9672 or Partridge D4A, size 6-10.
Thread: Orange.
Tail: Tan elk hair.
Aft hackle: Brown, palmered over abdomen.
Abdomen: Cream fur or synthetic.
Wing: Tan elk hair tied in front of abdomen.
Forehackle: Brown, palmered over thorax.
Thorax: Orange fur or synthetic.

There are few secrets to fishing this large imitation. Just plop it along the banks. And I am reminded to tell you that right along the banks is not six feet out, it is six inches out. Put it into those difficult places where you are almost sure other anglers are too timid to cast. If you guess right, you're going to get a splashy surprise. When naturals are in the air over riffles, put your fly out into open water. If fish refuse it, give it some movement. Coax the fish, as if you were the insect and you wanted to get eaten.

A more commonly accepted dressing for the salmonfly adult is the Improved Sofa Pillow. It is tied on a 3X long hook, size 4 to 10. The thread is brown, the tail orange-dyed deer hair. The body is orange fur with a brown hackle palmered over it. The wing is thick, of elk hair. A wide band of brown hackle in front of this finishes the fly.

If you don't like the fly that I like, try this one. You might like it better. So might the trout, where you fish for them.

There are two related species, so similar that you will have trouble telling them from the salmonfly. The first is *Pteronarcys princeps*. It generally hatches west of the Cascades and Sierras. The second is *Pteronarcys dorsata*. It is an Eastern species, but exists in the east-flowing rivers of Colorado and other Rocky Mountain states. Emergence dates are similar, and the flies listed will take trout when these insects are on the streams.

Golden Stones
Calineuria californica

Chapter 7 _____

It was late June. I'd taken a job teaching fly fishing every weekend on the Deschutes River. The drive from home took half a day. Then there were three days on the river, with the students. They were pleasant days, but they started at six in the morning, ended near midnight. After the long drive home, I'd have another draft of a videotape script to dash out before heading sleepily back to the river again for the next weekend's classes.

I got so worn down I needed a layover day on the river. It would be a busman's holiday, but at least I would get to fish myself.

I slept the morning of my day off nearly away. I lazed around reading and tying flies until early afternoon. The day was almost gone by the time I headed for the river. By then the sky was overcast, building toward a rain. The pickup bounced up the rough rock road that parallels the river above Maupin. I parked here and parked there, scanning the river with binoculars, looking for rising fish.

It didn't take long to spot them. They were sporadic, and close to the banks. I resisted the temptation to struggle down the scree-bank and fight the willows and briers along the river. But then I saw a rise, behind a boulder, that spelled size. The fish didn't come up again, but I knew its lie, and had a hunch I could coax it to the top.

On my awkward scramble down the steep slope I noticed lots of large stoneflies clambering about on grasses and sage stems. The salmonfly hatch was already over, so I collected one of these and had a look. It had tannish-brown wings and a flat brown head. I turned it over to look at its flip side; it was the rich yellow of a Golden Stone.

I worked into position to drop a fly behind the boulder. I broke my tippet and freshened it with a couple of feet that tested four pounds. Then I knotted on a big and bushy Stimulator dressing, with a palmering of grizzly hackle wound over a bright yellow body, beneath a deer hair wing.

The first cast was accurate. The fly danced on its tiptoes in the confused swirl of current below the boulder. There was a sudden detonation. I raised the rod, set the hook, and held on while I was nearly spooled. Some Deschutes River trout know how to use its powerful currents, some don't. This one did. The fight lasted a long time, and I thought the fish was large. But when I finally held it in my hand it weighed only a couple of pounds.

As the clouds built up, Golden Stones started to fly. They towered high into the air. Some descended to lay their eggs in the riffles. Some escaped the water to fly again, but many were taken in such violent splashes that I always thought the cause was a monstrous trout. But I fished the Stimulator through the rest of the afternoon and evening without taking a single fish that weighed more than three pounds.

Golden Stones cause that kind of activity during their June and July emergences. But it is always sporadic, and it is usually difficult to determine that they are the direct cause of your fishing success. They are, however, so very large that trout smack their imitations even if only a few of the adults are around.

The hatch starts in middle to late June in the coastal states, late June to early July in the Rocky Mountain states. It goes on for two to four weeks, so the period of their availability is long. On many rivers it overlaps with salmonfly importance.

The actual Golden Stone hatch takes place at night. The time of day that they are most important is evening, when the adults that emerged on previous nights get active. On cloudy days they tend to come out earlier in the afternoon. There are few consistent egg-depositing flights. They'll cloud the air one day, be totally absent the next. About the only reliable thing that can be said about them is that fish are greedy for them when they can get them.

Golden Stones are distributed throughout the West. Their preferred nymphal habitat is rushing water over cobble or stone bottoms. But they live in streams wherever the water is cool and entrains lots of oxygen. Like salmonflies, their numbers can be awesome on some of the larger rivers. Unlike salmonflies, Golden Stones are also abundant in smaller headwater streams.

The life cycle of the Golden Stone lasts two to three years. The nymphs are fiercely carnivorous. They prowl among riffle rocks, preying on midge and caddis and blackfly larvae, mayfly nymphs, and whatever else they can kill and eat. When ready to emerge they crawl out of the water. This almost always takes place after dark, and the nymphs often crawl far from the water. That is why few of their shucks are found on streamside stones.

After emergence the adults hang around in the vegetation for severald days. They mate in the foliage. They are not as awkward as salmonflies, but they do get into enough trouble along the banks to make a large dry fished there very interesting to some fair-sized trout. Their unpredictable evening egg-depositing flights are extremely gratifying if you happen to hit one just right.

To me, the most important stage of this stonefly is the adult. I know that will be considered heresy by many modern anglers. I might well be wrong. The nymphs do migrate to shore for their emergence. They are available for the long period of their two- and three-year life cycle. But I have simply not found evidence of trout feeding selectively on them, as I have on salmonfly nymphs.

Golden Stone Fly

The adult Golden Stone is an easy character to recognize. It is large, from an inch to nearly two inches long. It has a flat and characteristically golden-brown head. Its wings are held flat over the abdomen, and are light brown. Its antennae and tails are both long. Its body is golden-brown to a bright golden-yellow, brighter on the belly than the back.

The Stimulator is a fine pattern to fish when the Golden Stones are out. There are two reasons. First, it looks a lot like a Golden Stone, and trout eagerly take it for one. Second, it looks a lot like a lot of other things, and trout eagerly mistake it for them, too. That gives it a double-barrelled kind of energy: it's a good imitation, and a good attractor.

The Stimulator was originated by Randall Kaufmann, author of *American Nymph Fly Tying Manual* and *Lake Fishing With A Fly*. Randall is one of those world-traveling anglers who has tested the waters on most of the continents. It can be assumed that a fly that finds his favor takes trout in a wide variety of situations. The Stimulator is one of Randall's favorites. His dressing for it follows.

STIMULATOR
Hook: Mustad 9672 or Partridge H1A, size 6-12.
Thread: Orange.
Tail: Deer body hair.
Rib: Grizzly hackle, palmered.
Abdomen: Yellow fur.
Wing: Deer body hair.
Hackle: Grizzly, palmered through thorax.
Thorax: Orange fur.

The quality of the hackle will determine how well this fly floats. Use the very best you can find and the fly will walk the water on the tips of its toes. This might seem foolish, because a stonefly doesn't do that. But they do whir their wings when they get on the water, and all that variegation of grizzly hackle represents a lot of motion, even when the fly is standing still.

Tossing such large flies calls for at least moderately stout tackle. I arm myself with my 8½-foot rod for a no. 6 line. Leaders should be kept

strong, around 3X, or 4- to 6-pound test. When fishing the banks, with their confused currents, I grease the leader all the way to the fly, to keep it on top so the tippet doesn't sink and tug at the fly. Jim Schollmeyer taught me this when fishing caddis dries. I didn't believe him until I tried it myself. I always thought the leader should sink. There are times when it should, others when it shouldn't.

The Stimulator represents a certain amount of motion all by itself. But when Golden Stones oviposit over riffles, fish the fly with some twitches, or even skitter it across the water to entice the fish.

In an attempt to correct my heresy about the nymph, I'd like to say that fish do eat a lot of them. There are likely times when they take them selectively. There are more times when they don't.

I would like to propose that a Gold Ribbed Hare's Ear nymph, tied on a weighted size 4 to 8 hook, will take most fish that are feeding on them, whether selectively or not. Fish it with the same tactics used for salmonfly nymphs: dead drift, right down on the bottom. The Golden Stone nymph is more active than the salmonfly nymph, and I have read reports that nymphal patterns imitating it shoud be fished with a rhythmic teasing swing. If dead drifting along the bottom doesn't work, give this streamer tactic a try. It might cause a trout to whack it.

A related species that is similar to the Golden Stone is the Willow Fly, *Doroneuria pacifica*. Its populations overlap with Golden Stones. They are so similar that distinguishing them can be difficult, and is largely unnecessary, though Willow Flies do tend to be slightly darker.

I have collected Willow Flies in many Oregon, Idaho, and Montana streams. They are reported to exist in heavy populations in Colorado rivers such as the famous Frying Pan. The same patterns and presentations that work for Golden Stones work well when Willow Flies are on the wing.

Pale Morning Dun

Ephemerella inermis,
E. infrequens

_____ **Chapter 8**

The elusive Green Drake hatch let us down. Richard Bunse and I were on the Henry's Fork in late June, in an attempt to catch the peak of the West's most famous hatch, on the West's most famous river. But we missed it, or it never happened, or it happened while we glanced the other way. We never did find out what happened.

Still, we had excellent fishing. And we had it all to ourselves, despite the battalions of other anglers who had migrated to the Fork at the same time for the same reason.

I have this habit, when I get thwarted by a hatch, of going out early, to make sure I see what is going on through the whole day. Sometimes I wind up snoozing with the insects in streamside grasses. But other times I am rewarded by the discovery of a hatch that nobody else notices is going on. That's what happened on the Henry's Fork.

Henry's Fork in late June is a gorgeous place to snooze in the grass. The banks of the upper Railroad Ranch section are gently sloped, nibbled like pastureland, with a scattered bloom of yellow mule's ear flowers that stare downward like narcissi until the sun puts a finger under their chins and slowly lifts their faces upward. But they weren't peering up yet when the first fish began to rise. It wasn't even nine o'clock.

I galloped to get Bunse.

There were two selfish reasons for this. First, his fly boxes are a lot better stocked than mine, and he is generous with them. Second, he is a better fisherman than I am, and figures out what is going on faster than I do.

As soon as he got to the river, Bunse confirmed my first judgement. "Yep," he said, "fish are rising." Then he went to work.

Bunse at work is interesting to watch. He is stout. He wears a lot of different hats, and he wears them all well. That day his hat was a World War I Army drill sergeant's hat, brown, with a wide brim and rounded crown. When he is wadered, vested, hatted, and standing in the water with a fly rod in his hands, Bunse looks immensely sure of himself. He is.

I have two pictures of Bunse that day. The first is of him standing butt deep in the placid waters, slightly leaning to one side with his rod jammed under an arm, studying something in a cupped hand that dripped a stream of water. The second is of him gently lifting his rod to set the hook into a fish 40 feet below him and just off to one side.

The fly that fooled the fish was selected to match the insect Bunse had captured out of the current and held in his hand. There was little time between the pictures, and there weren't any pictures between. He figured it out that quickly.

I sidled up to him while he played the trout, a nice rainbow of about 16 inches. I got the necessary information—"It's a Pale Morning Dun!"—and accepted the offered fly, a size 18 Pale Morning Comparadun. I went away with it and found a rising trout of my own.

The one I found came consistently to the surface about a foot in front of a boulder in the current. This would not have been a problem, except that the fish rose right in the middle of the upwelling where the current lifted to split itself around the rock. That presented a problem because, as I analyze it with backward glances now, the upwelling stretched the surface film and straightened the leader out, no matter how much slack I tossed into the cast. That made it almost impossible to present the fly without making it drag just as it reached the critical point in its drift.

I explain all this so carefully so that you will appreciate how good I am when finally I boast that I hooked that trout. I did it this way: I cast and cast and cast until by accident one time everything worked out just right and the fish came up and took the fly, and I didn't even know what I had done differently. Still don't.

Bunse and I both took lots of trout that morning, and the next. We had the river all to ourselves from 9 o'clock until 11, when the hopeful in the Green Drake army risked their Seal-Dries crossing the pole fence at the upper end of the property, only to find that those great mayflies were absent again from the battlefield.

Fred Arbona, in his *Mayflies, the Angler, and the Trout* (Winchester Press, 1980), defies popular judgement and lists the Pale Morning Dun as the most important Western mayfly hatch. Most people put it beneath the Western Green Drake, which often overshadows it. But I agree with Arbona.

Pale Morning Duns begin to come off in late May here in Oregon. They are still coming off in late July. They start in early June in some low-elevation Rocky Mountain waters. The first hatches might be as late as early July at higher elevations. But they go on and on, lasting sporadically until October. Their hatch period is a lot longer than the Green Drake's, and they emerge in good numbers nearly every day.

Their daily hatch period is just as beneficial to the angler. At the early end of the seasonal hatch, they come off at midday, from around 11 o'clock until 1 or 2 o'clock. But as the hatch goes on, it moves up earlier in the morning, and they often come back to emerge some more later in the afternoon or evening. Sometimes a few of them are around all day, especially on cloudy days.

Pale Morning Duns are distributed everywhere in the West. They inhabit mountain streams in New Mexico. There are good hatches in waters on both sides of California's Sierras, Washington's and Oregon's Cascades. I have fished them throughout Idaho and Montana. Wyoming and Utah have heavy populations, and Colorado rivers have some of the best Pale Morning Dun hatches in the West.

Gentle currents like Alberta's Bow River are perfect for Pale Morning Dun nymphs. They prefer moderate to slow water, though there are good hatches in the peaceful stretches of almost all freestone streams. Spring creeks have them in what might be the highest densities.

These small mayflies have a one-year life cycle. They browse along the bottom until they are ready to emerge. Then they let go and swim toward the surface. A few inches from the top they cast the nymphal skin, and from there they struggle and are buoyed upward until they break through the surface film. Because its wings are wet, the dun rides the surface for a dangerous amount of time. When they finally take wing they head for land, where they rest until it's time to get the molting, mating, egg-laying, and dying done. This usually takes a day or two.

I believe that the inches-short transition between nymph and dun, just under the surface, might be the most important stage of the insect. But my experience with that stage is too sparse to make any conclusions. My fishing has been with dun dressings. In my own fishing I have had the most success with them, which reflects my prejudice for dry flies.

I have not encountered a fishable Pale Morning Dun spinner fall. From what I have read and heard, some people consider them important, others do not. I cannot report on them because I have not encountered them when it was advantageous to match them.

Pale Morning Dun

Pale Morning Duns are the Western counterpart of the Eastern Sulphurs. The Western duns are size 16 and 18. They have three tails. The bodies are a pale yellow with shades of olive, and it is a color that is always described as elusive. It is. The wings are pale-grayish with a note of pale yellow in them. Even the descriptions are elusive!

Fly pattern selection for the dun is simplified by the habitat preference of the nymph. Since they emerge in gentle currents, even on rivers that have

77

their rough stretches, flotation and visibility are not the first considerations. Proper silhouette is the most important aspect of an imitation.

A Pale Morning Comparadun captures both the body and wing outlines of the natural, and has nothing in the way to interrupt the trout's view of them. The dressing for it follows.

PALE MORNING COMPARADUN
Hook: Mustad 94833 or Partridge L3A, size 16-18.
Thread: Pale yellow.
Wing: Natural cream deer hair.
Tails: Ginger hackle fibers, split.
Body: Olive and pale yellow fur, mixed (Hareline no. 8 and no. 11).

It helps, when tying flies with bodies of any kind of fur, to apply floatant at the same time you twist the dubbing onto the thread. By binding it up with the fly as you tie it, the floatant gets plenty of time to dry. The result is a fly that floats a lot longer than one you dress when you prepare to cast it.

Tackle for fishing these small flies should be light, though not so light that you can't fight an occasional Western wind. I use a 5-weight rod, an 8-footer because that happens to be the one I like best. Most fly fishermen think any rod shorter than 9 feet is a waste of space above them, and I think that is fine, since there is little question that a 9-foot rod works at least as well as an 8-foot rod, probably better.

Leaders for this fine fishing should be between 12 and 15 feet long, tapered down to 5X or even 6X, though you will have trouble keeping Henry's Fork rainbows out of the weeds with anything less than about 3-pound test. Modern leader materials that are strong for their diameter are highly recommended when fishing this fine. A tippet that is three or four feet long will give you some extra float with playful mini-currents tugging at your fly on spring creek surfaces.

In most water inhabited by Pale Morning Duns, positioning yourself to cast downstream to feeding fish will give you a marked advantage. Upstream casts that let the fish see your line in the air, or your leader on the water, are likely to frighten fish.

I mentioned that my experience with the emerging nymph is too thin to base conclusions upon. But the dressing offered by Fred Arbona in his mayfly book seems so simple and so sensible that I'd like to present it here. I would also like to recommend you track down a copy of his book, since it says a lot that condenses sense out of things that contain confusion.

Mr. Arbona's dressing for the emerging nymph follows:

EMERGING PALE MORNING DUN
Hook: Mustad 94833 or Partridge L3A, size 14-18.

Thread: Light olive prewaxed 6/0.
Tails: Ginger hackle fibers.
Body: Pale yellow dubbing (Flyrite no. 25) ribbed
    with olive 6/0 thread.
Wings: Two hackles; four turns with a light dun
    hackle for wings; two turns with a light ginger
    to imitate the legs.

This is a wet fly dressing, or more properly, a
"flymph." It represents a stage of the insect that
the late Pete Hidy described as ". . .no longer a
nymph, but not yet a fly, hence a flymph." It
makes sense. Fished just under the surface, or
even in it, it should take trout during Pale Morning Dun hatches. Use dry fly gear. Fish the fly on a
dead-drift presentation to rising trout. Or try pulling it under the surface and letting it drift around
on a casual wet fly swing, gently led by the line
and leader.

You might find that it outfishes the dry. I'm going to try it next summer; I might discover the
same thing. If we do, we should let Mr. Arbona
know we appreciate him.

Salmon Fly

Langtry Special

Golden Stone

Stimulator

Pale Morning Dun

Pale Morning Compara-dun

Damselfly Nymph

Green Damsel

March Brown Dun

March Brown Parachute

Little Olive Dun

Little Olive Parachute

Speckle-Wing Quill Dun

Callibaetis Compara-dun

Green Drake Dun

Bunse's Natural Dun

Green Rock Worm

Green Caddis Larva

Gray Sedge Pupa

Partridge & Green

Gray Sedge

Deer Hair Caddis

Spotted Sedge

Elk Hair Caddis

Grasshopper

Letort Hopper

Trico Spinner

Trico Poly-wing Spinner

Fall Caddis

Fall Caddis

Midge Pupa

TDC Nymph

Green Damsel
Coenagrionidae

<hr />

Chapter 9

Dick Haward and I stood in his boat on a lake high in Oregon's Cascades. It was a calm late-June day. Irritating rays from the bright afternoon sun glanced from the still surface up into our eyes. The boat swung lazily on its anchor rope. But it didn't matter; there were rising fish in front of us no matter which way the boat turned.

It was frustrating fishing. We went without a hit for two hours. A few mayfly spinners hovered above the water, but there was no evidence that fish fed on them. Some caddis danced ten feet up in the air, apparently waiting for evening before descending to deposit their eggs. Fish weren't rising high enough to take them.

I peered into the water for a long time without seeing a single thing that might incite the minor riot that was going on all around the boat.

Dick is a retired executive. He is tall, lean, silver-haired, precise. He leaned over his side of the boat, pointed at the water, and said, "Dave, what's this?"

I looked. "It's a damselfly nymph!" I shouted, as if Dick were miles away. "So that's what the fish are taking."

The nymph was out of range of my short-handled aquarium net. We watched it paddle laboriously away from the boat, with a side-to-side wiggle that propelled it patiently along. Just after it had gone out of sight, about 15 feet from us, a brutal boil erupted right where it was heading when we last saw it.

That was the end of the insect, but it was the beginning of some interesting fishing for Dick and me.

We both tied on damselfly nymph dressings, cast them out, and began retrieving them with a variety of hand-twists, strips, and jerks of the rod tip. We never did discover precisely which was the most effective retrieve. Fish seemed to take at random. But our flies got startled often enough to keep us at least a little happy. Dick released one rainbow that weighed three pounds.

Damselflies get active in late May, build up in importance through their peak in June, and taper off through July. Though actual emergence from the water takes place at dusk or after dark, the nymphs are most available to trout while swimming toward shore in the afternoon and early evening.

Green Damsel nymphs are distributed throughout the entire West. They are found in lakes, ponds, and some very slow-flowing streams. They are most important in still water that is shallow and weed-filled. They are not important in moving water, nor very often in lakes that are deep and free of rooted vegetation.

Most damselflies have a one-year life cycle. The nymphs are predators. They lie cryptically along the stems of aquatic vegetation, depending on their stick-like form for camouflage. When a smaller organism ambles carelessly into range, the lower labium of the damsel shoots out like an extended arm, grabbing the prey. The jaw is then retracted, and the luckless victim is eaten at once.

The hidden life cycle of the nymph keeps it pretty well out of harm's way until time for emergence. Then it is betrayed by the necessity to crawl out of the water to complete the transformation from nymph to adult. Nymphs that live out in the lake must make the hazardous trip to shoreline reedbeds, downed logs, or lily pad flats. Many end the trip far from shore, in brutal swirls like the one Dick and I saw punctuating the life of the damselfly nymph that swam past his boat.

Such punctuation is common in the nymphal stage of the damsel. Those that complete the migration crawl out of the water on plant stems or half-submerged rocks and logs. The adult then slowly emerges from the nymphal skin, and within a few hours is ready to fly away. Because they are so susceptible to bird predation, nature usually protects damselflies by having this happen at night.

The adults mate in the air or on vegetation. The eggs are deposited in the stems of plants or along the submerged part of floating logs and other debris. It is common for the males to hold on to the females while oviposition takes place, continuing to grasp them behind the head, as they do when mating takes place.

Damselfly Nymph

To the fisherman, the most important stage of the damselfly is the nymph. Its importance is greatest during the time of its migration for emergence. These migrations are always difficult to discern. They are frustrating. Few of the nymphs swim by in easy sight. You've got to search for them. If you see just one or two swimming in open water, you should suspect this is what is going on.

Recognizing damselfly nymphs is easy. They have a slender and very elongated body. At the end of it are three caudal gills in place of tails. These are shaped like willow leaves, and have a network of veins that gather oxygen out of the water and deliver it to the cells of the insect. The thorax of the nymph is slightly thicker than the abdomen. The eyes of the natural are at the sides of the head, and are beadlike. The hinged labium of the nymph is tucked away under its head.

Damselfly nymphs have some of the aspects of the chameleon. They take on the coloration of the predominant vegetation on which they live. As a result they come in two primary color phases, green and dark brownish-olive. Many people prefer to dress patterns for both shades, but it is likely that a single green dressing will be sufficient in almost all cases in the West.

Polly Rosborough's damsel nymph dressing, as given in his book *Tying and Fishing the Fuzzy Nymphs* (Stackpole, 1978), is the dressing that produces most consistently for me. His dressing for it follows.

GREEN DAMSEL
Hook: Mustad 38941 or Partridge D4A, size 8-12.
Thread: Olive.
Tail: Light olive marabou fibers pinched off 3/8"
    to 1/2" long.
Body: Pale olive rabbit fur, Hareline no. 19.

Legs: Olive-dyed teal flank.
**Wingcase: Olive marabou, one shade darker than
   tail.**

The marabou in this dressing does an excellent
job of capturing the swimming motion of the
natural. This is a sinuous movement, and is hard
to imitate with any dressing that has a hard body
or a stiff-fibered tail and wingcase.

Tackle for fishing these flies should be at least
modestly powerful, since they are fairly large.
Long casts can be an advantage, and rods of 9 feet
or more, casting weight-forward lines of no. 6 or
7, do it best. Most of the time the nymph migration takes place up near the surface, just a few
inches deep. When this is the case, dry or inter-
mediate lines do a fine job. When the nymphs are
deeper, which you can only suspect, you will need
to probe with a Wet Tip, Wet Head, or full sink
line until you achieve the right depth.

Leaders should be 12 to 15 feet long, and should
be tapered to 3X or 4X, about 4- to 6-pound test.
The potential for a large trout exists whenever you

cast a damselfly dressing over Western waters. It is best to be prepared to land the largest trout the water might offer.

Presentation is based on the swimming movement of the natural. Short one-inch strips of line, or twitches with the rod tip, are most commonly recommended. Any retrieve prescription includes variety, and demands that pauses be inserted every few feet.

I have had most hits when the fly is left lifeless, with long pauses between short stripping retrieves. It is very difficult to detect such takes. A strike indicator is seldom used in lakes. I rarely use one myself. But I know that when the sun is on the water it is almost impossible to watch the tip of your line for the slight twitch or jump that marks a strike. An indicator improves the odds considerably.

Adult damselflies are reported to be important in some situations, especially when sharp wind gusts knock the adults out of reedbeds. My own experience with this is so thin that I will not offer a dressing for the adult damsel here.

Gray Sedges

Rhyacophila, 57 Western Species

_____ **Chapter 10**

Jim Schollmeyer and I got our signals mixed up, down in the gorge of the Santiam River, one bright July day.

Jim stands six-foot-three, so he lifted the rubber raft and descended the precarious trail to the river while I brought along the light stuff. We arrived at an eddy next to a chute that swept down to a short waterfall. The fast water was only about 20 feet wide, but we'd have to paddle across it fast or be whisked over the falls.

We constructed an elaborate plan of attack. Then we jumped into the raft, were caught by the current, and started paddling in circles as we shot down the chute.

Jim frowned at me, shook his paddle handle toward the far shore, and we made three concerted strokes in the same direction. I was on the downstream side of the raft; I peered over the lip of the falls just as we were swept safely away from it into a back eddy on the far side of the chute.

After we'd gathered our scattered emotions, we boulder-hopped downstream for a quarter mile. We idled the morning away exploring with a variety of nymphs and dries. But we didn't take many fish. We munched our noon sandwiches, watching out over the water. Soon Jim noticed something that I did not. Far out across the water, little splashy rises shot up, nearly invisible in all the activity of the vigorous river.

The Santiam was 200 feet across where we ate lunch. It was a fast run, with pockets of white water around lots of half-submerged boulders. Though the run itself was too deep and fast to wade, somebody Jim's size could work his way out to the edge of it. From there he could cast a fly to the rising trout.

While I sat on a high boulder and took pictures, Jim waded out and began casting over the fish. It took him about two changes of flies before he took his first trout. It was a 14-inch rainbow that tossed itself and lots of water around in the July sunshine. Jim brought it to his big hand, released it gently, and cast again. Over the next three hours I took six rolls of film, shooting Jim casting, Jim mending the drift of his fly, Jim setting the hook, Jim playing trout after trout.

When he finally waded back in, I asked him what he had been using. "There were lots of gray caddis dancing around out there," he answered. "I used my Deer Hair Caddis."

I had never heard of Jim's Deer Hair Caddis. I said, "Let's see it!" He handed me one. It was a bushy dry, size 12, modeled on Al Troth's Elk Hair Caddis. It had an olive body palmered full length with the best grade blue dun hackle. Its wing was natural deer hair, flared over the body of the fly. That was all there was to it.

"I use it whenever those grayish caddis are on the water," Jim told me.

"When is that?" I asked.

"Just about all summer," he answered.

Jim had worked out his dressing for the Deer Hair Caddis over a period of about three years. It was an attempt to match the many grayish caddis that he found bouncing around over Western riffles and runs. They were out there for long periods, almost every day, active on the kinds of sunny afternoons when other insects seemed to hatch briefly or not at all.

Jim's research, and the fly he concluded from it after three years, fit perfectly with all that is known about the *Rhyacophila* caddis.

Gray Sedges hatch from late May through the end of September, and even into early October. Their period of greatest importance extends from late June through the month of July on most rivers, though they might be regionally important at any time during their long hatch period.

The emergence itself is generally sporadic. But the adults are most active in afternoon and evening, ovipositing their eggs in the swiftness of riffles and runs. That is why Jim's dressing worked so well when midsummer sunshine struck the river. Even if the trout were not feeding selectively on Gray Sedges, they were used to seeing lots of them around.

Gray Sedges are distributed throughout the West, in fact across the entire continent. The preferred habitat of the larvae, called green rock worms, is fast water. They are found primarily in the runs and riffles of cool-water streams. Because the West is so loaded with their favorite kind of water, the Gray Sedges are important in all the Western states.

Their greatest abundance is in streams with a large percentage of tossed water. Logically, then, they are found in the heaviest populations of Sierra, Cascade, and Rocky Mountain streams with steep gradients.

The life cycle of the Gray Sedge lasts one year. The larvae prowl around among the rocks of riffles and runs. They are predaceous, feeding on midge and blackfly larvae, and the early instars of mayfly and stonefly nymphs that later might be larger than they are. They are unable to swim. Because of their fast-water habitat, they are often knocked loose by the current. They tumble hopelessly, waiting to be delivered back to the bottom or to the trout, at the whim of the current.

When ready to pupate they construct crude shelters on the bottom, out of pebbles and sand. Pupation takes about two to three weeks. When the transformation is complete they cut their way out of the shelter and rise to the surface. They swim fairly well, and are aided by buoyant gases trapped under the pupal skin. The ascent to the surface is fairly fast, but it is a time when the insects are very vulnerable to feeding fish. This commonly occurs in late afternoon or just at dusk.

The transition from pupa to adult takes place quickly. The pupa pops through the surface film, and the adult emerges and flies away almost instantly. Mating takes place in streamside vegetation, where the adults hang around for two to three weeks. Hatching is often sporadic. But the adults from each day's emergence accumulate, there are lots of them around, and there always seems to be quite a few of them bouncing around in the afternoon sun. They touch down to the water to wash away their eggs. Some swim under to lay them on rocks or logs along the bottom.

There is no single most important stage of the Gray Sedge. Larvae, pupae, and adults all have their moments of collective trouble with trout, and all cause trout to feed on them actively.

Recognition of the larvae is relatively easy. Their name implies two elements of that recognition: green rock worms are green, and they are collected from rocky bottoms in fast water. They range from half an inch to three-quarters of an inch long. They are slender, worm-like, and have

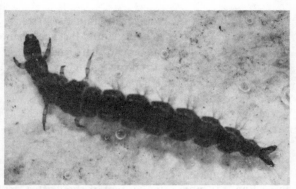

Green Rock Worm

two anal hooks where their tails ought to be. The head and the first segment behind it are hardened; the rest of the insect is soft and fleshy, poorly armored for such a combative beast.

Because they live in fast water, dressings for them should be weighted. The best pattern I have found is dressed on a curved hook to capture the typical posture of a tumbling natural. The dressing for it follows.

GREEN CADDIS LARVA
Hook: Mustad 37160 or Partridge K2B, size 8-16.
Thread: Tan.
Weight: 8-10 wraps of fine lead wire.
Body: Green fur or synthetic.
Legs: Gray partridge.
Thorax: Tan hare's mask fur.

91

Fish feeding in fast water will almost constantly be on the lookout for these caddis larvae. Their imitations should be fished dead-drift, right on the bottom. If the water is shallow enough, use a dry line, 8- to 10-foot leader, and a strike indicator to let you know when you have a take. Pickups will normally be subtle, not easily detected by feel. If the water is too deep or too fast for dry line fishing, switch to Hi-Density Wet Tip line, and shorten your leader to about four feet.

The best tackle includes a long rod and double-taper line, for maximum control of the drift. Cast at an angle upstream, give the fly plenty of time to sink to the bottom, then follow its drift with the tip of the rod. Fish it out until it is well below you, then pick it up and cast upstream again. Watch for the slightest twitch or hesitation in your line. Fish are unlikely to rap these larval imitations.

Gray Sedge (Pupa)

The elusive pupal stage is very difficult to detect. You simply won't see them unless you happen to catch a trout and examine its stomach contents. It is wise to suspect them whenever you fish a riffle or run in midsummer, see adult caddis in the air, but fail to take fish on dry flies. This deductive type of reasoning substracts what doesn't work from what is possible, until you wind up with what works even if you don't know why. It is an effective way of figuring things out in the evening when you can't see what is going on. And that is just when the pupal caddis is most important.

The pupal stage of the Gray Sedge contains the fully-developed adult within the pupal skin. It is difficult to identify this caddis pupa exactly without getting too scientific about it. Its body will be stout, sort of bulbous, green in color, size 10 to 16. The wingpads will be sloped back along the body, and will be greenish tan to greenish-gray, revealing faintly the color of the wings folded inside. The legs and antennae will be swept back along the body, prepared to be tossed by all the currents the pupa must brave on its way to the surface.

Imitations of this stage of the insect do not need to be exact, since trout feed on them in rough water. But it is important to capture the movement of the various appendages of the natural as it is tossed by the current. The best flies to do that are simple soft-hackles, written about in Sylvester Nemes's wonderful book, *The Soft-Hackled Fly*.

Nemes's dressing for my favorite Gray Sedge pupal pattern follows:

PARTRIDGE AND GREEN AND FUR THORAX
Hook: Mustad 3906 or Partridge G3A, size 10-16.
Thread: Green silk.
Body: Green silk floss.
Thorax: Black and brown hare's face fur, mixed.
Hackle: Gray partridge.

This simple dressing fishes effectively for the green caddis pupae for which it is listed here. But it also resembles a lot of other things, including drowned mayfly, caddisfly, and some stonefly adults.

Fishing soft-hackles calls for dry fly tackle. Lines should be in the no. 5 or 6 weight class. Leaders should be 8 to 10 feet long, tapered to 4X or 5X, testing two to four pounds. Rods should be long, and lines double-tapers, to help control the fly.

Tactics call for tending the drift to keep the fly moving freely downstream. Casts should be slightly upstream or downstream from straight across. Mends and tosses of slack should be used to keep the fly a few inches deep, and as free as possible from any pull by the line and leader. The fly should tumble along like a leaf in the drift, or it should be led gently around on a slow wet fly swing. It should not gallop so fast that no trout would mistake it for a natural. When soft-hackled flies are fished right, trout hit them with a wallop.

The adult Gray Sedge is an extremely active insect. It flies erratically over the water, touching down to lay its eggs, flitting into the air again, swirling around in confusion. Fish know they must take them quickly; that is why the takes are often splashy.

Gray Sedge

Identification of adult caddis is confusing. They
are all similar in shape, with their tent-like wings,
short fat bodies, and long antennae. They have no
tails. The Gray Sedge group includes around 60
Western species. They are characterized by
speckled wings that are gray to brownish-gray,
and bodies that are green to tannish-green. They
range in size from 10 down to 16. Sizes 12 and 14
are most common.

Jim Schollmeyer's Deer Hair Caddis captures
the shape and color of the naturals. It also stands
up on its bushy hackle, giving the perfect impression of all the movement of the real thing. His
dressing for it follows.

DEER HAIR CADDIS
**Hook: Mustad 94840 or Partridge L3A, size
12-16.
Thread: Olive.
Hackle: Blue dun, palmered.
Body: Dark olive fur, Hareline no. 26.
Wing: Natural dun deer hair.**

You should use the very best grade hackle you
can afford for this fly. It will float longer if you
dress it with floatant when you tie it. The kind of
water where you will fish it calls for all the float
you can get.

Tackle for fishing these dries is standard. A long rod will help you cast farther across broad runs. Five and 6 weight lines work perfectly. Leaders need be no longer than 10 feet, tapered no finer than 4X. Upstream casts will usually work well enough, as fish in fast water are not so easily spooked. There is no need for the fine presentations called for on spring creek currents, though of course the more careful you are, no matter what the situation, the more fish you will catch.

The Western angler who wants to catch more trout will want to carry imitations of all three stages of the Gray Sedge: larva, pupa, and adult. Only a person who restricts his fishing to still or pokey water will get through a season without wishing he had these dressings at hand.

Spotted Sedge

Hydropsyche, 25 Western Species

_____ **Chapter 11**

The August heat was almost unbearable. Dad and I fished through most of the day with little luck. There were so many caddis along the river that clouds of them erupted whenever we swished past a sagebrush clump. But the insects were not active, and few of them were getting onto the water where trout could get at them.

In late afternoon thunder clouds began to stand high over the Deschutes River canyon. At 5 o'clock they burst, and we fled for the tent. We huddled inside while lightning played in the rims above us. Thunder claps had the flat sharp whaps of incoming artillery landing too near you. It was two hours before the summer storm moved on, leaving the tent rain-struck and wind-battered.

We had dinner. Dad decided to stay and clean up in the aftermath of the storm. I took my rod and walked over the slick wet earth down to a rocky point that split a riffle. Where the point probed out into the fast water, it left an eddy on the downstream side.

There was less than an hour of daylight left. I didn't expect any fishing. But trout were rising at the edge of the eddy when I reached the river.

Something was going on, and the water nearly boiled. It was one of those hintless situations. There were no insects in the air. I had no idea what the trout were taking. But even these kinds of situations contain their own hints. In the past, when I've not know what is happening, a small March Brown Spider soft-hackle has often saved the situation. It is the same fly I use to imitate the emergers when March Brown mayflies hatch.

Because the predominant insect I'd seen all day was a small Spotted Sedge, I chose a size 16 March Brown Spider with hopes it might imitate the pupa of the caddis. When it's evening, and I don't know what is going on, I suspect caddis pupae, and tie on a soft-hackle.

Approval was intant. The first cast was short, from the rocky point out into the first few feet of riffle. The fly tumbled downstream. When it swept out of the fast water and into the upper end of the eddy the line tip shot forward and I raised the rod. A rainbow came up and danced around until I led it into the shallows. It was a fat one-pound fish. I rapped it and tucked it into the back pocket of my vest: breakfast for Dam and me.

I cast again. When the fly swung into the eddy, a fish hit again. A few minutes later I released a trout a little larger than the first. It went on that way for three-quarters of an hour. On almost every cast I saw the line tip twitch, or felt a subtle tap-tap. The fish were all between one and two pounds, and all fought hard. I suspected my tippet was getting worn, but I'm not the kind of fisherman who has the patience to stop and change it when the fishing is hot and daylight dwindles.

I made what turned out to be a final cast. There was the same soft take, but this fish felt heavy. It turned at once out of the eddy and ran into the riffle. Then it turned its tail to the current and went swiftly downstream, with all the power of the river behind it. The line was all out, and about half of the backing, before the rod simply went dead. I reeled up and ran my fingers over the end of the tippet. It felt rough and frayed.

It was too dark to tie on a new fly, and I had no flashlight. I turned around and started to walk off the point. A voice called to me quietly out of the darkness. "You were doing pretty well out there."

"Yes!" I answered. I was startled. I peered and discovered a young man sitting on a rock. He was wadered, with a fly rod across his knees.

"What were you using?" he asked.

"A soft-hackle," I answered. "A little brown one. It imitates the pupae of those caddis you see around all day."

"You're using bamboo," he said. I carried a Leonard I'd bought when I couldn't afford it. It was in those days when bamboo rods were rare along Western streams.

"Yes," I answered, and we got into a long conversation. He introduced himself as Randy Stetzer. He said he had been sitting behind me all the time I'd been catching fish. I concentrated so intently on the fish that I did not hear him arrive.

"I was just enjoying watching you fish," he told me as we sat there talking in the darkness. "You sure looked like you knew what you were doing." It was a flattering conversation, and I left the river half an hour later feeling like a true expert. Randy and I have been friends ever since, and he has gone on to become a famous Western tyer. He tied the flies for the color plates in my book *American Fly Tying Manual*.

Back at the tent, I got out a flashlight and cleaned the single fish I had kept. I poured a little water into a pickle jar lid, then squeezed the stomach contents of the fish into it. When I swizzled it around, what separated out were the remains of dozens of adult Spotted Sedges.

"The storm must have knocked thousands of them into the water," Dad said.

"The fish weren't taking pupae at all," I admitted. But I've never told the truth to Randy Stetzer, who was so sure I knew what I was doing that day.

The fishing I had in the aftermath of that Deschutes River storm seems to have been an accident. But there are so many Spotted Sedges around Western waters that they get into trouble with trout in all sorts of different ways. Gary LaFontaine, author of the fine book *Caddisflies* (Winchester Press, 1981), has called them the single most important trout-stream insect. I agree with him. There are more than 25 Western species that make up the Spotted Sedge complex, all in the genus *Hydropsyche*. Their numbers are almost unbelievable on streams that have favorable habitat.

Spotted Sedges begin to emerge in June. But the peak of their importance is July and August. They continue to emerge through September, and can be important at times throughout that entire period.

Daily emergence tends to be sporadic, with a day-long hatching cycle that keeps trout alert and feeding without cease. Oviposition tends to take place in the afternoon and evening. Such perpetual activity, when added to their great numbers, contributes to the importance of Spotted Sedges.

Wide distribution is another factor. The larvae require moving water, but live in most Western streams. Runs and riffles are their favored habitat. They are most abundant in rivers that have dams, rivers that are enriched by organic effluents, or rivers that receive nutrients from irrigation systems.

The life history of this caddis makes it most important where man has changed the nature of a river. The larvae feed by building tiny nets. These trap particulate matter adrift on the current. The larvae come out to browse on the nets periodically, then retreat into crude shelters to await the capture of more food. When man enriches a stream it increases the particulate drift. Filter-feeding Spotted Sedge larvae are just the fellows to take advantage of such a situation.

Dams stop the flow of a stream and allow the rapid growth of tiny planktonic organisms, which are then carried into the stream below the dam. In tailwaters, populations of *Hydropsyche* literally explode to take advantage of this new form of feed. This is not always a healthy situation, since the diversity of the stream is usually narrowed. It increases the importance of certain caddis larvae, but sometimes decreases the health of the entire fishery. Not always, though; tailwater fisheries are some of the best.

When preparing for emergence, the larvae seal themselves into their shelters, pupating inside. When the transition is complete, in a few weeks, the pupae cut their way out of the shelters, then drift along the bottom for a considerable period before ascending to the surface. Some drift again for a time just beneath the surface, before punching through it and emerging rapidly.

The adults fly immediately to streamside vegetation. Mating takes place there. The adults live two to three weeks. The hatch might be sporadic each day, but each day's hatch accumulates until there are millions of Spotted Sedges along the stream.

Adult Spotted Sedges are day-time insects. They bat around in the afternoon air, or lay their eggs in the evening. Most dive into the water and swim down to place their eggs on the bottom. This is remarkably risky behavior and trout are not unaware of it.

The life cycle of the Spotted Sedge makes it important to trout in all stages. Larvae are taken along the bottom, pupae as they drift or rise to the surface, adults when they land on the water and dive under it to lay their eggs.

Spotted Sedge Larvae

The larvae are worm-like, very similar to Gray Sedge larvae, from one-quarter to one inch long. They tend to be tan, though there are some green species. They have three hardened plates behind the head. There are rows of tufted gills along the abdomen, and usually tufts of gills at the base of the anal hooks. Their crude shelters, cobbled together out of leaves, sticks, and pebbles, are another identifying characteristic, seen if you collect your specimens by lifting stones out of the water.

Imitation is simply a matter of varying the color of the dressing given for Gray Sedge larvae, since their shape is the same. My favorite dressing for them follows.

TAN CADDIS LARVA
Hook: Mustad 37160 or Partridge K2B, size 8-16.
Thread: Brown.
Weight: 8-10 turns of fine lead wire.
Body: Tan fur or synthetic.
Legs: Brown partridge soft-hackle fibers.
Thorax: Brown hare's mask fur.

This dressing imitates the most common Spotted Sedge larvae. If you encounter green ones, which you likely will if you fish much, simply imitate them with the Green Caddis Larva given in the last chapter.

Tackle and techniques are exactly the same for Spotted Sedge larvae as for their relatives, the Gray Sedge larvae. Use dry lines, long leaders, and a strike indicator. Cast upstream or up and across, and let the weighted nymphs tumble down along the bottom. Set the hook at any hesitation, as takes will usually be gentle. If the water is deeper, use a Wet Tip line and shorter leader to present the fly in the same manner.

Spotted Sedge pupae have the same bulbous bodies, slanted wingcases, and trailing legs and antennae that characterize Gray Sedge pupae. Then tend to be tan, but many species are green. Imitation calls for the simple soft-hackle that I used the night thunder struck the Deschutes. The dressing for it follows.

Spotted Sedge Pupae

MARCH BROWN SPIDER
Hook: Mustad 3906 or Partridge G3A, size 10-16.
Thread: Orange.
Rib: Narrow gold tinsel.
Body: Mixed fur from hare's face.
Hackle: Brown partridge.

This is another pattern from Sylvester Nemes's soft-hackle classic. It should be tied sparsely to imitate the natural caddis pupa. Tackle and tactics are similar to those used for Gray Sedge pupae. Long rods, double-taper no. 5 and 6 lines, and leaders 8 to 10 feet long work best. Cast up and across the stream, and let the fly drift slowly down

with the current. When it has reached the end of its free drift, coax it gently across below you, leading it with the rod and line. This type of presentation imitates both the rising natural, and the drifting natural before it pops through the surface film.

The adult Spotted Sedge is characterized by its speckled wings. The body is usually a brown or tannish color, though sometimes they are olive. The wings tend to tans and browns, though again some species have coloration similar to the Gray Sedges. There is considerable shading throughout the caddis order, which is fortunate for us since it keeps fish from getting overly selective.

Because their habitat is moving water, usually runs and riffles, the best dressing for Spotted Sedge adults is well hackled, and is suggestive rather than imitative. By far the most popular and most effective pattern is Al Troth's Elk Hair Caddis. His dressing for it follows.

Spotted Sedge

ELK HAIR CADDIS
**Hook: Mustad 94840 or Partridge L3A, size
  12-18.**
Thread: Tan.
Rib: Fine gold wire.
Body: Tan fur or synthetic.
Hackle: Ginger, palmered.
Wing: Tan elk hair.

When tying this fly, wrap the body, then palmer the hackle back over it. Catch the hackle under the wire rib, so that it is permanently secured to the fly. When it is tied properly, the Elk Hair Caddis will take dozens of fish without falling apart.

Tackle is standard for dry flies. Number 5 and 6 rods are fine. Leaders should be 10 to 12 feet long, tapered to 4X or 5X, 2- to 4-pound test. Presentation should usually be upstream or across stream, in runs and riffles. When fishing Spotted Sedge imitations there is seldom need for the downstream, slack-line tactics called for on spring creek currents.

There are, however, populations of these insects on some Western spring creeks, such as Henry's Fork of the Snake. When these are encountered, and the fish are fussy, you might try clipping the bottom hackles from the Elk Hair, allowing the body to float flush in the surface film. This increases the imitative qualities of the flies. Many professional tyers whose clientele fish on smooth waters tie their size 16 and smaller caddis dressings without any hackle at all. It makes sense, but only if you intend to fish slick currents, and no others.

It is wise to recall that Spotted Sedge adults dive under water to deposit their eggs. Many are taken by trout. The perfect imitation is the same March Brown Spider that represents the rising pupal stage. Fish it with the same tackle and tactics, and you should do very well.

This soft-hackled wet fly imitates two stages of the same insect. It should be the first thing you reach for when you notice Spotted Sedge activity, but you can't entice fish to the dry fly.

107

Grasshoppers

Chapter 12 ─────────────────

As I waded through the tall dry grass toward the river, a wave of grasshoppers tumbled along in front of me. So many clattered their wings that I kept hopping into the air myself, thinking each sound was the warning waggled tail of an angry rattlesnake. It was August; the air was hot and windless in the Deschutes River canyon. By the time I got to the bank I was nearly melted into a couple of puddles at the bottoms of my wader legs.

It felt good to slip into the water. I just sat there with my legs in the water and let the cold river restore me. After a while I stood up, removed my fly from its keeper, and began wading and fishing upstream.

It is my favorite way to fish rivers like the Deschutes. The banks drop off steeply, the current is swift almost right against the shore. But there is a narrow zone where one can wade with caution, using the bankside hand for balance, grabbing willow branches or bunchgrass clumps. But I watch what I grab; I don't want it to rattle its tail and grab me back.

I worked along very slowly, fighting the current but enjoying its coolness. I cast the large and bushy dry fly short, tight against the grasses that overhang the river. Most casts landed with a smack. The fish that rose to it were swift and sure. They took hard, ran far out into the strong river, jumped high and fought well. I brought a dozen to my hand in a couple of hours. I'd waded less than a hundred yards.

Hopper time is like that. It is almost like fishing a tiny stream: you move along slowly, popping a dry fly to all the prospective holding lies: into pockets and indentations in the shoreline, above and below rocks and boulders in the nearby current. It is not at all like fishing a hatch. In some ways the unpredictability of slashing grasshopper takes makes it more exciting.

Grasshopper time begins in late July, peaks with the heat of August, trails into September so long as the weather remains dry and the days stay warm. Even after nightly freezes the ungainly insects will slowly unlimber their stiff legs and begin to hop feebly when the sun finally strikes them. They are most active when the sun is hottest. Scorching afternoons are prime hopper hours. They are hours when few other insects are out, making them all the more important.

Grasshopper

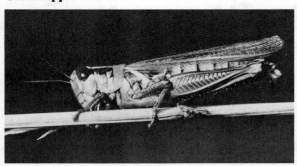

Grasshoppers are distributed throughout the West, throughout the world for that matter. They are most abundant where the climate is dry. They are not important along the rain forest rivers of the northwest coast. They are very important on the rivers draining the Sierras, the east side of the Cascades, and the entire Rocky Mountain region. Meadow streams and sage-lined rivers usually have hoards of them.

Because their preferred habitat is land, hoppers are most important along the edges of rivers. This transition line between the terrestrial and aquatic environments is a line they cross uncomfortably and unwillingly, but far too often to suit themselves. They cross it most often when a strong dryland wind beats and bends the grass stems. They are awfully unhappy when they land in the water. They kick and struggle in futile disgust.

Trout wait eagerly along the same transition line. When hoppers helicopter in after crossing it, trout take them brutally.

Pattern rationale for grasshoppers is based on the kind of water over which you will fish them. Most of my own experience is on rivers with strong currents and at least slightly rough surfaces, which is another way of saying average Western rivers. I have not found it necessary to imitate the insects exactly when fishing this kind of water.

I confessed in the first chapter that I was a slightly reluctant tyer. If I can get out of tying a couple of dozen dressings to imitate a specific insect, I'm glad. A lot of this is because I no longer have the time to tie that I had in the past. Another lot of it is that my vest has begun to bulge in recent years, and part of what I seek is simplicity and a slimmer profile.

I'd like to confess in this chapter that I rarely tie flies to imitate grasshoppers. The reason is simple: most of the time other dressings will do. Which ones? Not surprisingly, the useful Elk Hair Caddis, in its tan dressing, takes most of my trout during hopper time. That is what I used on the hot

August afternoon described in the early paragraphs of these grasshopper notes. It was tied on a 2X long, 2X fine size 12 hook.

Another excellent pattern is the Stimulator, given as the dressing for the adult Golden Stone. When tied on a size 10, 12 or even a 14 2X or 3X long hook, the Stimulator will usually be the only fly you need during grasshopper time.

A Muddler, dressed with a floatant, is also an excellent hopper imitation. Its size and silhouette are very similar to the natural. A greased Muddler floats low in the film and gives an excellent impression of a disgusted and struggling grasshopper on the water.

When I do find need for a specific grasshopper pattern, it is always on a meadow stream or spring creek with a slick surface. Because hopper populations are great on this kind of water, it is worth bulging the vest a little to make room for a few flies that match them. I have found that the Letort Hopper works well, even though it was originated as an Eastern dressing. It is an accurate representation of the natural, which makes it an excellent fly to use on smooth water. The dressing for it follows.

LETORT HOPPER
Hook: Mustad 9671 or Partridge H1A, size 8-14.
Thread: Yellow.
Body: Yellow fur or synthetic.
Underwing: Mottled turkey quill.
Overwing: Deer body hair.
Head: Spun and clipped butts of overwing.

Many Western grasshoppers have yellow bodies, and are perfectly represented by the Letort Hopper as it is listed. Others have shades of olive in them, and some are nearly green. It is possible

111

that you will want to vary the body color to match the color of the hoppers prominent in your region.

Tackle for hopper fishing is not unlike standard dry fly gear, though you might find it advantageous to use a no. 6 or 7 dry line when grasshopper winds blow. Leaders should be 10 to 12 feet long, tapered to 3X or 4X, 4- to 6-pound test. Keep them stout enough to withstand the type of takes you'll get with these large flies.

Presentation should violate some of the rules of dry fly practice. First, it is fine to smack your fly to the water. That's the way the naturals land, and a properly splatted fly can attract the trout's attion. Second, if trout ignore a dead-drifted fly, give it some twitches. If possible, inveigle some disgust into its float. Remember, a grasshopper is not happy when it discovers it is on water.

Ants

_____ **Chapter 13**

I t was hot that June. I was in summer school at
Oregon State University, taking catch-up
classes that I had missed during the regular ses-
sions. It felt good to toss felt-soled hiking boots
into the car, grab the little Leonard, and flee the
hot campus for the cool coastal hills.

There was only one way to reach the small forested stream. I parked the car a half mile almost
straight above it, then skidded down on deer and
elk trails until the streambed itself stopped my des-
cent. The ascent, hours later, would be agony. But
nobody else ever fished there, so the fishing was
always worth it.

Because I preferred to fish upstream, with dry
flies, I hiked down the stream for a couple of
miles. I would eat lunch before turning around
and fishing carefully back up. I had to wade in
places going down, so I dabbed a dry fly here and
there as I went. It was difficult to fish the tiny
stream that way, but I usually picked up a few fish
before I arrived at my lunch spot. But this day I
rose nothing, and I was surprised by it.

Halfway down I switched over to a wet Alder,
and not long later finally hooked a trout. Since a

113

fish dinner was a treat during those poor college days, I killed it and cleaned it and creeled it. Then I examined its stomach contents.

The trout bulged with black ants. They were large, about a size 12. Most were wingless, but I noticed a lot of unattached wings in the sample, and concluded that all or at least most had been winged when taken by the trout.

I searched the streambanks then, and finally found a winged black ant crawling around on a rock. On an impulse I flicked it into the stream. It struggled a moment, then sank. A subsurface wink a few feet downstream was incriminating evidence to the murder I had caused. A little careful snooping revealed a lot more ants.

I switched from the Alder to a Flying Ant imitation. I did not dress it with floatant, preferring instead to fish it wet around the rocks in the short riffles and runs of the little stream. The fish began to hit in all their normal lies. It was not the easiest way to fish a wet, straight downstream, but whenever I could get a decent drift and swing I would get a rather insistent tug.

I finally reached the place where by tradition I sat with my back against a mossy tree trunk to eat lunch. I tore chunks out of a loaf of bread, to eat with slices of cheese and summer sausage, while I waited for a beer to cool in the shallows. Alongside me, in a row of ferns, were six dark native cutthroats, 12 to 14 inches long. All were taken on the ant dressings fished wet. I didn't catch that many on the way back upstream.

Mating flights of carpenter ants take place fairly early in summer. But there are so many species of ants that they are present on streams, and even lakes, from May through the end of October. Their season of greatest importance is when the weather is warmest, because that is when they are most active.

Ant

When winged ants are out they can get into trouble anywhere along a stream. On lakes, they are often blown to the water far from shore. I have been frustrated for hours out in the center of a 15-acre lake, where I never would have suspected ants. When finally I peered over the side of the boat and examined the water closely, I could see them struggling feebly in the film. I switched to the same dressing I used on the small stream, but in a much smaller size. I dressed it lightly so it would float flush. The fish thought it was fine.

Worker ants are wingless, and more abundant than winged ants. They scramble around in the shrubs, sagebrush, and grass alongside most rivers. Though they can be important at all times, they are most likely to prompt active feeding when a hot wind precipitates lots of them to the water. When this happens the best fishing is right next to the banks.

Ants have waxy bodies. The largest of them sink readily, because of their high density compared to their relatively small surface area. But the smallest ants float in the surface film until they reach a riffle, or until a trout takes them under. Fly pattern

115

selection should take into account the shape of their bodies, the lack of excess appendages, and their nature that gets them taken by fish sometimes dry, sometimes wet.

Though it is possible to tie a series of dressings that imitate the insect wet and dry, winged and wingless, large and small, I find that one dressing can be used for all if it is fished correctly. The dressing for my favorite ant pattern follows.

FLYING BLACK ANT
Hook: Mustad 94840 or Partridge L3A, size 10-26.
Thread: Black.
Abdomen: Distinct ball of black fur.
Wings: Blue dun hackle points.
Legs: Sparse black hackle.
Thorax: Distinct ball of black fur.

The dubbing should be divided into distinct segments, with a gap between. The hackle should be sparse. On the smallest sizes, from 18 down, leave off the wings. They are not needed; neither is the agony of tying them on a fly so small.

Many ants are brown. They are called Cinammon Ants, and you might want to tie the above dressing with brown fur and brown hackle, though you also might want to wait until you encounter a situation that calls for them. Often, fish will take the black version when brown ants are on the water. Size seems to be most critical, followed by shape and finally color.

If you want to fish the fly dry, dress it with floatant. If you want it to sink, leave it undressed, and soak it in saliva so it will get through the surface film quickly. To make the fly float flush, clip the top and bottom hackle fibers before fishing it. If the ant you imitate is wingless, snip the wings from your fly.

Tackle should generally be fine when fishing ants. The smallest of them are most important on smooth currents, so you should use light lines, long leaders, and tippets of 5X, 6X, and sometimes 7X, as fine as one pound test. For the kind of forested stream I fished during my college days, when trout took the ants wet, leaders the length of the rod, tapered to 4X, worked fine.

Presentation of ant dressings varies with the situation. I have fished them wet, with fairly coarse tackle, on the smallest mountain streams. I have also seen the time when they had to be fished with careful casts to highly selective fish in clear and calm water. When that happens, it is best to work into position upstream from feeding fish, so you can present the fly gently downstream to it, right in the feeding lane.

On lakes, use long and fine leaders. If fish rise steadily, cast out and allow the fly to rest, waiting for cruising fish to come to it. If few fish are rising, try to pattern an individual, and cast ahead of it, where you think it will be ready to rise next.

Trout feeding in lakes are often erratic. But when you get the rhythm and direction of a fish figured out, and place a tiny dry ant dressing in front of it, the reward will be worth all the puzzlement and frustration.

Beetles

_____ **Chapter 14**

It was early September, but the weather was
still August-hot. A high wind blew through the
pines above the spring creek, swaying their tops,
twisting and swishing their limbs. All through the
afternoon trout kept rising almost imperceptibly,
and I kept trying to catch them on an exhaustive
array of dry flies and subsurface nymphs. Nothing
worked.

It was Fall River, in Oregon's Cascades. The
rule on this frustrating stream is to start with large
flies, say size 20's, and work down from there until you find what trout will take. On this day, I
went all the way down to size 28, but still the trout
tilted up their noses at my flies while taking something else all around them.

In my fly fishing infancy, it always took me a
long time to remember the aquarium net folded into an inside pocket of my vest. Even when I
remembered it I seldom used it: how could the currents hold something that I couldn't see, right in
front of my eyes? But this time I dipped it to the
water, and the act of doing so forced me to stoop
over. When I did that, I could see the surface held

some creatures so small the only thing that gave them away was the slight disturbances they made in the surface film.

I captured a couple in the net. They were beetles, black and tiny. I looked up at the swaying trees and had little trouble puzzling out their origins.

I selected the smallest black dry fly I had, a midge tied on a size 20 hook. I used my clipperrs to amputate its tails, wings, and hackles. All that was left was its fur body. I extended my leader to 7X, then tied on this tiny lump of fur. It looked to be about a size larger than the beetles I'd plucked from the water. I rubbed some floatant into it, then cast it over the nearest rising fish.

I don't know if the fly was too large for the selective trout, or if my presentation was youthfully flawed, but I managed to coax only a handful of rises in the next three hours. Few were solid takes. Even fewer became fish on. Fewer yet became fish caught. But I had some action, while I had no action at all before I collected the beetles and did my best to imitate them.

Beetle

Terrestrial beetles begin to get active in late May. They go on through fall, their numbers dwindling only after a few hard frosts have driven them to their hibernation and eventual death. If there is a peak of importance to fishermen, it is when summer winds waft them into streams. Daily cycles are similar: beetles are most important when afternoon winds shudder the trees.

Though they are distributed everywhere, beetles are perhaps most important where forests creep down to the water's edge. It does not matter if the water is a lake, pond, river, or creek. Wherever beetles land on the water, trout are glad to find them.

There are too many species to attempt imitating all beetles. There are big ones and small ones, fat ones and skinny ones, black ones, yellow ones, red ones, and all sorts of other ones. There are 30,000 species. You are not likely to run into all of them, and if you do you might not want to tie flies to match each of them.

Fly pattern rationale for beetles is based on trying to condense some sense out of the astonishing array of them. The best way to do that is with hope: tie one beetle dressing and hope the ones you run into look a little like it. The pattern I have found to look most like the widest variety of beetles is a simple deer hair tie. If nothing else, it doesn't take long to tie a size range of them, and it doesn't take much room to carry them. Here is the dressing for it.

BLACK CROWE BEETLE
Hook: Mustad 94833 or Partridge L3A, size 14-26.
Thread: Black.
Legs: Butts of a few of the shellback hairs.
Body: Thread windings over shellback hairs.
Shellback: Black-dyed deer hair.

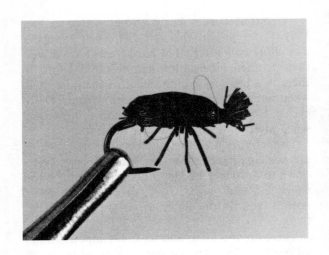

To tie this simple dressing, start by tying a small clump of deer hair to the hook shank, tips to the back. Leave the butts long. Draw three of them out to each side and trim the rest. Wrap a body of thread over the hair. Now draw the shellback over and tie it down at the head. Clip off the excess fibers and you're finished. It doesn't take long to tie a dozen. Dress them with floatant when you tie them, so that when you fish them they will float.

These beetle dressings float flush in the surface film. It is almost impossible to see them. Half the time you won't know whether they are awash or sunk. But fish each cast out with great patience and hope. If you see a rise in the area of your fly, raise the rod tip gently. You might just be setting the hook.

Both tackle and tactics for fishing these flies should be as fine as any you own. Since beetles tend to be most important on glassy water like Fall River, you will need to lengthen your tippets and hone your stalking and casting skills. Since yours are already better than mine, I won't give you any more advice.

Tricos
Tricorythodes minutus

_____ **Chapter 15**

The bull bellowed early, sometime in the second week I was on the river, waking me up just as the first morning light crept into the cottonwood flat. I was not ready to leave the warmth of the down bag, but I decided it would be a good idea to see what the river looked like at dawn.

It was the first year after I got out of the Army. I was driving across country, leaving the East, heading for home and the hunting seasons of fall. I arrived in Montana in September, with three weeks to kill. After diligent map reading I found a wheel-rut road leading off across a sagebrush sidehill, then down toward the Big Hole River valley far in the distance. I opened a gate, drove through, closed the gate behind me, and bounced down the dirt road.

I opened and closed several more gates. None were marked "No Tresassing", so I kept driving toward the river. I expected to be turned back by a sign at any instant; I expected the road to peter out far from the river. But neither expectation came true, and finally I nosed the pickup into a cottonwood grove right down in the river bottom, five miles beyond the first gate.

It was a lonely and beautiful place to camp. The ground was like a park, its grass cropped and green. It was too early for the cottonwoods to turn their fall colors, but the smallest maples and alders beneath them had already lost some September leaves. Every time the wind gusted through the grove these crackled and sprang into the air like flocks of startled birds.

My only company was an owl that hooted through the chilly nights, and a massive whiteface bull that spent the warm days browsing the sage hills, its nights nearby in the cottonwoods. The bull had a burnt-reddish hide that gleamed when the sun lit it. Its heavy horns arced far out, then back in to point at each other in front of its curly white face. The bull had to lower its head to look at me over its horns as I traipsed in my waders through the trees on my way to and from the river. At first I thought it might be about to charge. But it didn't, and I didn't run, and we became sort of standoffish friends.

The morning the bull bellowed early, I agonized out of the bag and into wool long johns. I pulled cold jeans over the woolies, then my waders over the jeans. My rod was always left strung, on top of the camper, so all I had to do was reach for it and stumble groggily off through the trees to the river. The owl hooted a staccato question at me: "What're ya doin' up so early? What're ya doin' up so early? What're ya doin' up so early?" It wouldn't leave the question alone; soon it had me wondering, too.

The lowest reach of the Big Hole River has gentle pools set apart by long vigorous runs. I had spent the first week on the river fishing Muddler Minnows and large Gray Nymphs, casting as far as I could and letting them swing on long arcs across the runs. I caught plenty of fish that way, some of them over 3 pounds. I ignored the pools, though I suspected they might hold the largest trout. I ignored the mornings, too, because I had fishing that was good enough for me all through the lazy afternoons and evenings.

When I reached the river that cool dawn I was not surprised by any activity on the water. I walked upstream along the broad gravel bars for a mile or so, to warm up. When the first rays of the sun struck my back, I sat on a fallen cottonwood and rested for a while. I decided to turn around and fish a Muddler back down toward camp and breakfast. Perhaps the streamer would move an early trout.

It didn't, but something else did.

As I fished down I faced the sun. It slanted in at a harsh angle, just over the low hills. Above the long pool just upstream from camp, the sun ignited a million dancing specks of light. It was impossible to tell what they were. Nothing rose in the pool, but I decided to sit and watch a while.

About 9 o'clock the sparks suddenly went out, and as suddenly the entire pool was marked by tiny rings. I glanced at the sky to see if it had just begun to rain. It hadn't. These were rises.

I had never heard of Tricos in those early days. Nor did I carry a collecting net. The relationship between insects and rising trout was only beginning to clarify itself in my mind. I waded out and cast my Muddler, dragging it through the rises. That didn't work very well. It left a trail vacant of rise rings in its wake.

I went down in size, longer and finer in tippet, as the feeding went on. But a size 16 was tiny to me then, and that's as far down as my fly boxes went. I fished for two hours, but finally gave up in frustration. It never did occur to me to examine the water, to see what the fish might be taking. I waded out of the river and headed across the gravel bar toward the camp cottonwood grove.

I was hungry. I was irritated at the trout that still rose in the pool behind me. I looked down at my waders and saw dozens of tiny dark specks on my legs. I didn't know what they were. Some kind of insignificant bug. I flicked them off my waders with a forefinger as I walked angrily toward camp.

Since that frustrating day I've learned a lot more about Trico spinner falls. These tiny mayflies hatch on Western rivers as early as the middle of June. They continue through August, peaking in importance in early October, and trout continue to feed on them as long as the hatches are heavy.

Their early morning emergence tends to conceal Tricos from many anglers. Fishermen reluctant to leave the warm bag, those who linger over third cups of coffee, might never know Tricos exist on a favorite stream. The hatch might come off as early as daylight, lasting until 7 o'clock or so. Most often it begins between 8:30 and 9:30, lasting until 11 or as late as noon.

Tricos are distributed throughout the West. Their habitat preference is for streams with slow to moderate currents. Their populations are greatest over bottoms that are silty, and in spring creeks thick with weed and algal growth. Streams like Henry's Fork of the Snake in Idaho, and Fall River in California, have heavy populations. But Tricos are also prolific in the slower stretches of freestone rivers like Montana's Big Hole River.

There are sometimes two generations of Tricos in a single season on the same river. The first comes off in June and July, the second in August and September. Where they are abundant, their morning hatches can go on daily for two or three months.

Most of the life history of this insect is spent in the nymphal stage. They are herbivorous, browsing along the bottom or among trailing algae and weeds. Emergence takes place in one of several ways. Some hatch along the bottom and the dun makes its way to the surface. Some swim to the surface as nymphs and the dun emerges there. Some crawl out on rocks, and the dun emerges alongside the stream.

The transformation from dun to spinner takes place within a few minutes to an hour or so. The molt has been reported to happen in the air in some cases, but this has been difficult to confirm. The spinners mate in the air, almost immediately, and the males die soon after. They fall to the water in masses, and this accounts for the sudden winking out of all those early morning aerial lights above the Big Hole River.

The females deposit their eggs over a longer span of time, but the entire adult life of the Trico mayfly lasts only two to seven hours.

Trico Spinner

The spinner is the most important stage of this insect. Spinners fall to the water in great numbers, and cause selective feeding despite their tiny size. The nymphs are certainly taken by trout. But it is difficult to determine when it is happening, and it is unlikely that trout are often fixed on them. The duns are also taken, sometimes selectively, but not as regularly as the spinners.

Recognizing Trico spinners is not difficult. The first key characteristic is their size. Large specimens are size 20; small ones are size 26. They have three tails, those of the female about the length of the insect's body, those of the male about three times that length. The wings are clear, usually lying spent on the water. Unlike most mayflies, Tricos have no hind wings.

The body of the male is dark brownish-black. The female has a pale olive abdomen and a brown thorax. There is some question which is more important to the angler, the male or the female. I have yet to resolve it for myself, so suspect that the best program is to offer alternate dressings that cover both. Only the body color need be changed, and it is not certain that color is the most important aspect of the fly pattern. A fly the right size but the wrong color is more likely to take fish than one the right color but the wrong size.

But it won't hurt you to get both the right color and the right size, if you can determine which your trout prefer. My favorite Trico pattern follows, with alternate body materials listed.

TRICO POLY-WING SPINNER
Hook: Mustad 94833 or Partridge K1A, size 20-26.
Thread: White midge thread.
Tails: White hackle fibers, split.
Abodomen: Olive or dark brown fur.

**Wings: White polypro yarn, tied sparse and spent.
Thorax: Dark brown fur.**

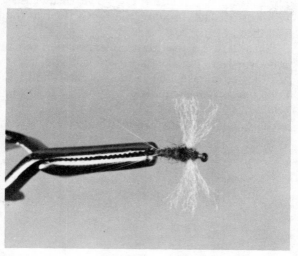

Tying such small flies might seem difficult at first. That's because it is. But once your fingers get used to the fine proportions, and the gentleness needed when working with fragile hooks and threads, it becomes easy. The tying itself sometimes becomes the challenge, and it is easy to forget to go out in the morning and go fishing with them.

I keep my Trico patterns in a 12-tablet Anacin tin, in order not to lose them in with larger flies. Cut thin patches of packing foam to fit both the top and bottom of the tin. Double-stick tape holds the patches firmly in place. A tin holds about three dozen tiny flies.

Tackle for fishing size 20 to 26 flies should be very fine. A soft rod keeps you from popping fragile tippets. Number 3 or 4 double-taper lines, or weight-forwards with long front tapers, are best for gentle presentations. Leaders should be 12 to 15 feet long, and should be tapered down to three feet of 7X or even 8X tippet. Tricos are almost always important in the kind of water that requires stealth and subtlety. Use downstream casts, with lots of slack to allow the fly to float to the fish ahead of line and leader.

You will lose a few trout on leaders that test a pound or less. But if you experiment with these tiny flies and fine techniques during early-morning Trico hatches, you will land a few trout that you would not otherwise be able to interest. Most important, you will extend your fishing day by a few hours.

Fall Caddis
Dicosmoecus sp.

_____ **Chapter 16**

I n October, on the Deschutes River, not many
people think about trout. Summer steelhead
hold in the long, waist-deep runs. People line up
along the edges of them, casting long, letting the
fly come around in a coaxing swing, taking a step
or two, casting again. And again, and again, and
again, until finally there is that satisfying heavy
thud and quick surge and leap that is the inception
of battle against a hot summer fish that brings its
strength in from the ocean.

I enjoy this madness, too. When the alarm goes
off in the tent at dawn I resent its insistence, but I
cannot ignore it. My favorite riffle is out there,
talking excitedly before it tapers off into a choppy
run. I know exactly where the steelhead hold in it,
and I know exactly where the take will come, if it
comes.

One recent October morning I shut off the
alarm and pulled on my waders. Mist rose from
the riffle when I arrived. Nobody else was there.
The riffle and its attendant run are too small for
true steelheaders to argue over. But for me it is
perfect. I can fish it in an hour, change flies and

run through it again, then head back in for breakfast. I do not race other fishermen; I race the sun. When the sun looks over the canyon wall, steelhead will no longer move up to take flies.

When I swished through the sage, descending down the bank to fish the riffle, I noticed lots of large caddis clinging to stems. Some dropped off, letting go their holds by instinct, when I brushed too close. There were more caddis in the wheatgrass and willows that grew along the water's edge.

I did not imitate the caddis to catch the steelhead. I fished the run first with a Green Butt Skunk, tied low-water style on a size 4 hook. On mornings when the Skunk fails I go through again with a bright fly tied on a smaller hook. But the Skunk didn't fail that morning.

The take always comes when I am standing at the corner, right where the riffle becomes a run. My fly, swinging on its arc after a 50-foot cast, is two-thirds of the way back to my side of the run when the fish hits, quick and sharp, then ejects itself from the water to tumble end over end in surprise. Half an hour after the first strike I did not land the first fish. I landed a second that took in the same place.

Before the morning was over I released a third steelhead to find its way out of the shallows and back into the waist-deep water of the run. Then the sun struck, and fishing that was too good was suddenly over.

I sat in the grass at the edge of the river, my wadered feet in the water. The sun felt good on my cold legs. I watched the caddis limber up. A few of them began to explore around in the grasses and sage. But I knew it would be hours before they started the day's flight: hours in which I could have breakfast, tie some flies, take a snooze, make some notes on the morning's fishing, and read a chapter in a favorite book.

Fall Caddis are afternoon and evening insects.

The sun made its graceful arc, and went behind the opposite canyon wall, before I set my book aside and, like the caddis, began to limber up. The morning was for steelhead, the evening for trout.

I headed for a riffle again, but not the steelhead riffle. What I wanted was a riffle with a special kind of corner. I wanted it to break over from fast water to slow right next to the bank. And I wanted the eddy at the edge of the riffle to be hidden far back under the alder trees. There aren't many riffles like that on the Deschutes, but I know where they are, and I know they are worth finding when Fall Caddis are out.

Working in to the right position was hard. I had to wade deep in water made treacherous by large boulders. I had to bend over until my nose almost touched the water, to get back under the low-swept alders. Once I was there I had to wait like a heron while things settled down again. Finally I was ready for the cast: a mere flick that had to be perfect or it was alders I'd catch.

Flick! The fly rode the water fine, going in circles on the eddy under the trees, as if puzzled at where it was. A shadow appeared under it, riding deep, uncommitted. I twitched the fly, and waited . . .and whap!

There are only three eddies like it within walking distance of camp. I came back in late, stepping in the circle of light thrown by my flashlight. The three fish I released in the evening were not a third the weight of the three I released in the morning. Not one was over 17 inches long, a couple of pounds or a bit more. But the satisfaction in taking the trout was nearly as large as that of taking the summer steelhead.

Fall Caddis emerge in late September and early

October in Western rivers from New Mexico to Alberta, from California to British Columbia. They are most abundant along the Pacific Coast states, and considered to be of prime importance in summer steelhead rivers. But I consider them to be more valuable to the trout fisherman, though dry dressings that match them will take the larger anadromous fish.

These insects are so large that they draw up the largest fish. Lunkers that are normally reluctant to feed on top drop their reluctance when they get a chance at something the size of a Fall Caddis.

The preferred habitat of the larval stage of the Fall Caddis is medium to large rivers with strong currents and clean stone bottoms. They are not abundant where the water is sluggish. They are not important where the bottom is silted or weedy.

The life cycle lasts one year. The larvae live among the bottom stones, building protective cases of sand and tiny pebbles. They are large. The debris you often find in trout stomachs is usually undigested remains from Fall Caddis cases.

When it is ready to pupate the larva attaches its case to a rock. Sometimes colonies of them gather in the same area, often after migrating to the shallows. When pupation is complete, the insect cuts its way out of the case and makes its way to the surface. Again, much of this activity takes place at the margins of the stream, where trout cannot take advantage of it. Some Fall Caddis, however, do emerge out in open water.

The adults hang around in streamside vegetation for several days, sometimes up to three weeks. Mating takes place on the foliage. Most flying activity takes place in late afternoon or evening. Some deposit their eggs at the margins, along rocks and logs that are half submerged. But many fly out over the water, dipping the abdomen to the river, struggling to cast the egg mass loose.

134

In my experience, the egg-depositing adult is the most important stage of the insect. There are no doubt situations in which the pupae rise to the surface in open water, and trout would no doubt be eager for them if they did. But I have not been on a stream when it happened, and most emergences I have seen have taken place in such marginal water that trout were not a factor.

The adults are most important in water that has overhanging vegetation. That is why I fish eddies where it is necessary to work back beneath the brush. But fish will take them when they lay their eggs in riffles, too, and a large imitation fished there will often bring up fish that would not rise to smaller flies.

Fall Caddis Dicosmoecus

Recognition of Fall Caddis adults need not be scientific. The insects are large, usually a size 6 or 8. The wings are held tent-like over a fat body, which is yellowish-orange on the underside. The wings are tannish-brown, with bold black venation. The antennae are approximately body length.

Fly pattern rationale is based on the size of the insect, and the need to represent its movement on the water. They struggle and flutter, refusing to be still. Many times they bounce up and down, touching the abdomen to the water briefly, attempting to wash away the eggs.

The best pattern I have found incorporates lots of hackle to represent this motion. The dressing for it follows.

FALL CADDIS
Hook: Mustad 9671 or Partridge H1A, size 6-10.
Thread: Black.
Body hackle: Brown, one size undersized.
Body: Orange seal fur, or synthetic.
Wing: Natural brown deer hair.
Hackle: Two to 3 brown saddle hackles.

The hackle used should be of the best quality. It should float the fly up high above the water. With its excellent flotation, the fly can be twitched or even skittered across the surface. It is excellent when the natural caddis bounce and swerve and touch down occasionally to deposit their eggs.

The standard Bucktail Caddis, a Pacific Northwest tie with years of tradition behind it, is also an excellent dressing when Fall Caddis are out. This tie has a yellow yarn body, ginger palmered hackle, and brown deer hair wing. The difference is in the yellow body, and the sparser hackle. It does not float quite so well, but still it takes a lot of trout.

Another dressing that I find works well for me is the standard Elk Hair Caddis tied on a size 10 long shank hook. Though it is far from an exact imitation, it seems the impression it gives is good enough. I take at least half of my fish during Fall Caddis hatches with an Elk Hair dressing.

Tackle for fishing these large flies should be reasonably strong. I use a 9-foot rod with a no. 6 weight forward floating line. Not surprisingly, it is the same outfit I use for summer steelhead. I detest the necessity to carry two rods just because I am going after two kinds of fish in the same day.

Leaders need not be long; 8 to 10 feet is fine. Tippets should be no finer than 3X or 4X, about 5-pound-test being just right. You will get hung up in the alder branches, and you will want to be able to yank your flies out if you have to. And the fish you hook will often be large. It's best not to have to gentle them to the point of exhaustion before you release them.

Presentation should be soft when fishing eddies. But you can whack these flies onto riffles and it will only make the fish madder. No matter what kind of water you toss Fall Caddis dressings to, a few twitches might draw up trout that are reluctant to take a dead-drifted fly.

If you choose to fish a Fall Caddis dressing for summer steelhead, cast it down and across stream. Then skate it around as if it were a wet fly on a downstream swing. The wake it leaves will help you keep track of it. When the water boils all around it, drop your rod tip, look away across the river, and try to think of something else until you feel the fish pull.

If your nerves are anything like mine, you'll yank the fly away from the fish. That's okay; cast again. Steelhead will sometimes come to a dry fly five or six times before finally killing it.

Scuds

Gammarus and Hyallela

Chapter 17 _____

T here wasn't a cloud in the sky at dawn. The
desert wind bit. Rick Hafele had to pound
the sides of the plastic water jug with his fists, to
break up the ice, before he could get it to sur-
render water for our morning coffee. When we
saddled up for the long hike down the river we
wore wool gloves and stocking caps. We must
have looked like rotund bears, the way we were
bundled in thick down vests beneath our heavy
windbreakers.

It was late November, but this river was open to
fishing all year-round. It held rainbow trout that
grew to good size, protected by the simple formula
of distance and difficulty. The river was a long
drive from anywhere; once you got there it was a
long walk to the best water.

Rick and I hiked for three hours before we
strung our rods. An hour later we got together for
lunch, and wondered to each other why we had
bothered to string them. After lunch we split up
again, Rick fishing downriver ahead of me.

I lingered behind to look at a long, drowsy pool
that curved on its outside against the base of a
scree slope that had fallen away to the water. A

tumble of submerged rocks lined the far side, a long cast across the pool. From my position on the bank above the pool I could see that it was about five feet deep, that its current was choked with tendrils of aquatic weed growth, and that there were mysterious brilliant winks among the rocks along the opposite bank.

I dropped to the river and exploded a flock of chukars that were apparently down to drink. They sent my adrenaline rushing. I watched them sail on set wings across the river, land on the scree slope, then start hopping scoldingly up it. Barnyard roosters sound like muted chickadees when compared to the noise unhappy chukars can create. All the nearby hillsides echoed their protests for an hour.

I spent that hour trying to seduce those winks under water. I had a little luck. It came about this way.

First, I tossed a Muddler across and swept it back to me with all sorts of retrieves. None of them worked. Second, I sailed a Gray Nymph over, let it sink down deep, then coaxed it back toward me very slowly. Nothing winked at it. Third, I tried a tiny Pheasant Tail, knowing there were lots of little mayfly nymphs around, hoping it might look like one. It might have, but not to these trout. Fourth, and finally, I gave up and waded out to sit on the bank.

While idling unhappily in the sage, listening to the chukars still giving me hell, I glanced at the toes of my waders. They were coated with the weeds in which I had been wading. Something strange was happening: the weeds were wiggling and squirming. I checked it out.

I lifted a tendril of weed off a boot. I examined it closely and found that it contained almost a dozen olive scuds. They were so close to the color of the vegetation that they were almost invisible.

"If there are so many of them that they nibble at my toes," I thought, "there must be enough out there for trout to nibble at them."

I frisked my fly boxes, but found nothing closer in color to the scuds than a peacock-bodied Zug Bug nymph. As I have had to do so often in the past, I pinched this part and that part off the Zug Bug until I got something that slightly resembled the naturals. I tied it on, waded in, cast it out.

Had I gotten a chance, I would have brought it in with a hand-twist retrieve. But a trout interrupted my intentions, taking the nymph while it was still sinking down toward the weedbeds.

I took five trout from the pool in the next half hour. They averaged over two pounds; the largest weighed more than three. I might have been able to catch some more, but it seemed more important to me, at the moment, to catch up with Rick. It was urgent I tell him the news. It would be a rare chance to brag to him.

When I skidded to a halt beside him a little bit later, he was standing in a pool, calm as could be, playing a trout that outweighed any of mine.

"What did you catch it on?" I asked as he bent to release it. He didn't answer. He just plucked the fly from the jaw of the fish and held it out for me to examine. It was a Zug Bug from which not a single fiber had been excised.

"Scuds are thick in these weeds," he told me. "Have you noticed?"

Scuds are easy to not notice. They are not insects, and have no aerial stage. In order to discover them, one has to poke underwater. There are more scientific ways to do it than with your toes. Rick found them by running an aquarium net through the weeds. That is why, by the time I rushed down to tell him what I had learned, he had learned it long enough before to be releasing a four-pound trout.

Scuds are crustaceans, related to crayfish. They are most important in lakes, but are also found wherever moving water is slow enough to allow the growth of rooted vegetation. They have no "emergence" period, since they do not emerge. They are out there all year-round, feeding, swimming about, getting fed upon.

If there is a season of most importance for them, it is whenever aquatic insect hatches have dwindled. When other food is scarce, trout turn to scuds. That is why they were so important when Rick and I fished that desert river in November. Nothing else was going on, and fish had turned to feeding on scuds.

Scuds are distributed throughout the West. Their importance is greatest geographically wherever lake fishing takes precedence over streams. High Sierra, Cascade, and Rocky Mountain lakes almost all have good populations. The Kamloops region of British Columbia is perhaps the West's premier lake fishery, and scuds are a prime reason for the quality of the Kamloops rainbow trout.

Scud

Vegetation is the preferred habitat of scuds. As a result, they are most often found in the shallows, where weed growth is greatest. Penetration of sunlight filters out after about 20 feet in most lakes. Deeper than that there is little plant growth, and few scuds.

Recognition of scuds is simple. They have hardened segments the entire length of the back. A series of swimmer legs and real legs lines the underside, which when they swim is usually the upper side. They range in size from 1/16 inch to a full inch long. When collected they typically curl up, sometimes into a complete circle. But when they swim free in the water they straighten out, and are more stick-like than curved. Their tiny legs are an absolute whir of motion. They are short on rudder; their swimming is rather directionless.

Scuds range in color from olive to cream to grays and browns. When preserved the colors change dramatically, so never try to imitate a scud that has been pickled in alcohol. It won't look like anything the fish have ever seen.

Dressings for scuds should be kept simple. Although I often prefer something as elemental as the useful Zug Bug, an imitation that incorporates a shellback is more effective on lakes, where trout can be more selective than they are in moving water. One of the best dressings follows.

OLIVE SCUD
Hook: Mustad 7957B or Partridge G3A, size 8-16.
Thread: Olive.
Tail: Olive hackle fibers.
Shellback: Clear plastic from freezer bag.
Rib: Olive thread.
Body: Olive-gray seal fur mixed with olive rabbit fur.

Legs: Olive hackle fibers.
Antennae: Wood duck flank fibers.

It would be honest of me to confess that when I tie this fly, I usually omit the tail, legs, and antennae. The body fur, when picked out with a dubbing needle, does a fine job of representing the various appendages of the natural. I prefer the active impression over an exact but less lively imitation. I like the way the fur fibers vibrate in the water.

Though a case can be made for tying the above dressing in a variety of colors, scuds usually take on the color of the vegetation they live in. That is most often green, and I have found that to be the most important color in most Western waters. The most important sizes, in my experience, range from size 10 to 14.

It is tempting to tie the fly around the bend of the hook, because of the curve of the collected natural. But remember that when swimming, scuds are straight. It is always best to imitate a natural in the posture it has when fish are most likely to see it.

143

Tackle for fishing scud imitations calls for dry lines so long as you intend to fish them within three or four feet of the surface. Beyond that, it is best to switch to a Wet Tip line. If you need to go deeper than about eight to ten feet, a Wet Head line does it better.

The key to fishing scuds in lakes is to get the fly as near as you can to weedbeds. This calls for a countdown presentation, ticking off the seconds on each succesive cast until the fly comes back with weeds trailing from it. Then shorten the count a bit and you will be fishing the right depth.

Retrieves should vary. Scuds dart around erratically. A slow stripping retrieve sometimes works. A series of short, fast strips, with frequent pauses, also works. A hand-twist retrieve might be the most effective of all. Try them all.

The wise angler, when probing water where he cannot observe the trout, will experiment with depth and retrieve until a combination is reached that pleases the trout. They will let you know, but the takes will be subtle. Watch your line tip.

In streams, scud imitations should be fished near the bottom, or just above rooted vegetation. A dead drift works well at times. A standard wet fly swing causes the fly to swim across the current. But the line should be tended so the swing is as slow as you can possibly make it.

Again, takes will be subtle, and you should keep a close watch on the tip of your line for any hint of a strike.

Midges

Chironomidae

_____ **Chapter 18**

From my office windows in town I can look out across the Columbia River. Some days a dozen tankers cut slowly up the river or down, on their way to Portland or back toward the ocean and the Orient. The winter sun doesn't often strike my north-facing windows. But when it slants in and ignites the white bridgework of the ships, I know it is also beginning to slant in and warm the waters of a tiny pond in the hills just outside of town.

Most Oregon lakes are open to fishing all year. I seldom find them productive, during a typical wet and stormy winter. But this has been a mild February, and this is a different story.

A ship thrummed upriver at 11 o'clock yesterday morning. I wandered to the window to watch it. Sunshine reflected off the ship's tall masts. It lit white the backs of gulls that soared above the ship's wake, causing their wings to flash as they dove to fight over smelt and suckers stunned by the turbulent propeller. I opened the window and stepped out onto the roof of the building, where I was not supposed to go. I lifted up my nose and sniffed the air like an old coyote. It smelled warm.

I leaped back through the window, slammed the office door for a day on the writing of this book, and headed for the pond in the hills.

I didn't expect any action. It was still winter. The season wouldn't start for three weeks or so, with the first March Brown hatch. But there are worse places than a small pond to sit on a log and eat a lunch. And an hour spent practice casting over a bleak pond in February never hurts when real fishing starts in March.

But the pond wasn't bleak.

It would be poetic to say that puzzling rises dimpled its surface. But it wouldn't be fair: the rises that dimpled its surface weren't that puzzling. I've escaped to this pond before. When the first warming days of late winter arrive, the first hatch of tiny midges arises.

They are almost black, size 16. Early wrens and sparrows hop along the edges of logs, picking off a few stranded adults. The sparse trout population of the pond works patiently, taking the pupae just before they reach the surface. Because there are few fish, I work patiently, too.

Yesterday afternoon I caught only three. But they were the first three trout of the season, and they started the season a month before it normally starts. So they were very important fish.

That's what midge hatches do most often: they emerge so steadily through the year that they become important hatches when what we consider to be important hatches don't emerge at all. A small knowledge of these tiny insects often allows you to take trout at times when it seems they cannot be taken. It is true in late fall, when most hatches have ended. It is true in late winter, before spring hatches start. It is even true in mid-summer, when days that are too hot turn other hatches abruptly off.

Midge hatches are heaviest between April and

October, with a peak in July and August. They are distributed throughout the West, in fact the world. There is very little water that does not contain them. The best populations are in lakes and ponds with detritus and marl bottoms. Slow-flowing streams with soft silt bottoms also harbor great numbers of them. They are found in the slower sections of fast rivers, and often cause trout to feed selectively in situations where they are doubly frustrating because the angler never suspects a midge as the cause of the problem.

The life cycle of the midge seldom lasts a year. Most species are multi-generational, with two to five hatch cycles occurring each season on the same water. That is why there are midge hatches going on almost all the time.

Midge larvae live on the bottom or down among it. Some build small burrows, others worm freely around in the muck. They are largely detritus feeders, though some species eat whatever organisms come their way.

When ready for pupation, most midges build a rough shelter. They pupate inside, then escape the shelter when pupation is complete. The rise to the surface is slightly propelled by a pair of swimming paddles at the posterior of the pupa. Gases trapped under the skin also aid the ascent. But in most cases the rise is slow, the insect vulnerable to trout.

The surface film is quite a barrier to tiny insects. Midges often hang for a few seconds to a few minutes before they are able to penetrate it. This is perhaps their single most vulnerable time. Trout feed on them greedily whenever large numbers of them are suspended from the surface.

Once the surface film is broken, the pupal skin splits along the back and the adult emerges through the film. This takes time, especially if the weather is cold. Many of the insects get stuck in the shuck at this point. Again, fish feed on them heavily before emergence is complete and the adults fly away.

Adult midges are not readily available to fish, and are not very important to the fisherman. Though many larvae are taken by trout, there is no great evidence that they are taken selectively, either. It is the helpless pupa, rising to the surface, hanging in the film, or in the transition to the adult, that is the most important stage of the midge.

Midge Pupa

Recognition of midge pupae is not difficult, but there are closely related insects with which they can be confused. In general, if they are so close in appearance that you cannot tell them apart, the same patterns will work and you need not worry about it. Midge pupae have short, two-lobed swimming paddles. They have numerous fine respiratory filaments on the first segment behind the head. Their bodies are long, slender, and slightly tapered. The wingcases, legs, antennae, and head are all a darkish lump at the anterior end of the insect.

Midge pupae are generally small. But in northern latitudes, where there might be only one generation per year, they reach sizes matched with size 8 and 10 hooks. Most are matched with size 14 through 22. Colors range almost through the entire spectrum; if you collect diligently you will find cream, red, brown, black, and olive midges. In my own collecting, I have found that dark specimens predominate in fishing situations.

Fly pattern rationale for midge pupae is based on their simple shape, and their most common color. The best pattern I have found for them is the TDC Nymph, created by biologist Richard B. Thompson. TDC is an acronym for Thompson's Delectable Chironomid. The dressing for it follows.

TDC NYMPH
Hook: Mustad 3906B or Partridge L3A, size 8-22.
Thread: Black.
Rib: Fine silver tinsel or wire.
Body: Black fur or wool.
Thorax: Same as body, thicker.
Collar: White or cream ostrich herl.

I spent several years working out a dressing for a local midge hatch. The natural was so dark it was nearly black. Its body segments were distinct. Its thorax was a little lump at the head end. It had a tuft of gill filaments that it wore like a wig. It took me a few seasons of frustration before I arrived at a successful dressing.

One day while thumbing through a pattern book I discovered that the fly I had devised was a size 16 TDC. I was only 20 years behind its originator.

The dressing as listed is the most valuable I have found. But there is no denying that midges come in all sorts of colors. You should feel free to change the materials to suit your local hatches.

Tackle for fishing midge dressings should be as fine as you can get it without conceding the game to the winds that blow over Western lakes. Long rods that cast no. 4, 5, or 6 lines work well. Weight forward tapers are usually best for lake fishing, since they allow you to cast farther with fewer backcasts. Leaders should be long and fine, with extra tippet added to absorb the shock when you set the hook.

There are two tactics that work well with midge pupal patterns. The first is to use a 12- to 15-foot leader and a strike indicator. Dress the leader with floatant to within a few inches of the fly. Then cast out and work it back very slowly. The idea is to let the fly hang just below the surface. This represents the pupa just before it emerges.

The second tactic is to use a longer leader, and slightly weighted fly. Cast long. Let the fly sink patiently. Then bring it in with a retrieve that raises it slowly toward the surface. This represents the pupa as it rises toward the top for emergence. Leaders as long as 20 to 25 feet are used by experts at this method.

When the midge is in transition from pupa to

adult, it struggles with the surface film. A dressing that represents the insect in this difficult moment is the Griffith's Gnat. It doesn't look at all like the natural, but fish take it consistently during midge hatches, especially on rivers, and I would not want to be without it. The dressing for it follows.

GRIFFITH'S GNAT
Hook: Mustad 94833 or Partridge L3A, size 16-22.
Thread: Black.
Hackle: Grizzly, palmered.
Body: Peacock herl.

There have been times when this pattern solved seemingly unsolvable hatches. In recent years it has become one of those flies I reach for in desperation, when I've already run through all the patterns that seem to make sense. It must give the impression of a midge half in and half out of the surface film. It works exceptionally well in spring creeks, and on the smooth flats of freestone rivers.

Since these types of water are excellent midge habitat, I now try the Griffith's Gnat whenever invisibles are hatching and I can't determine what the trout are taking.

Conclusion:
Innovation

The patterns listed in the preceding chapters will fill a couple of fly boxes. They will take most of a pleasant winter's tying.

Fly tying is always the most fun when you have a specific goal in mind. When you plunk the last of a dozen Stimulators into a compartment, you will know you have the Golden Stone hatch covered. If you tie the dressings listed in this book, when you close the lid on the last box you will know you can fish the entire season on your home waters, or travel to any river or lake throughout the West, and not be surprised very often by a hatch that you cannot match.

There might be times when the dressings I have listed will not be the best for you. If spring creeks are your pleasure, you will want to tie flies that lean toward the more exact spectrum of imitation. As I mentioned in the opening chapter, I like flies that work well in all the sorts of water where I might encounter a specific species. If you can eliminate one of the water types, you might do better by selecting a pattern that works best exactly where you will fish it.

There are slight color variations within group-ings of species, sometimes even within a single species, from water to water. A Pale Morning Dun in Western Oregon is a size 16 with a pale yellow body that has some shades of olive in it. The same insect in Eastern Idaho is a size 18, and has an olive body with some shades of pale yellow in it. The difference is subtle, but when fish are ultra-selective it might be necessary to dress a fly with a more exact fur color.

You will always be a better fisherman if you are prepared to collect a specimen and match it with the best dressing you can find or tie. If you carry the flies listed in this book, you will almost always find dressings in your fly boxes that match the in-sects you will collect on Western waters.

But no *Western Streamside Guide* would be complete if it didn't advise you to be prepared to find an insect that I have not listed, or one that nobody else has encountered, created a fly pattern to match, and written about. This book would not be complete if it failed to warn you to be prepared to do some innovation of your own.

HATCH CHART

Natural	Artificial	Emergence Period			Emergence Time
		Coast	Rockies		
Western March Brown	Nymph: March Brown Flymph	March-May	April-June		11:00 a.m. — 3:00 p.m.
	Dun: March Brown Parachute				
Little Olive	Emerger: Little Olive Floating Nymph	March-Nov.	April-Nov.		11:00 a.m. — 4:00 p.m.
	Dun: Little Olive Parachute				
Speckle-Wing Quill	Nymph: Hare's Ear Wet Fly	May-Sept.	June-Sept.		10:00 a.m. — 3:00 p.m.
	Dun: Callibaetis Comparadun				

Western Green Drake	Dun: Green Paradrake or Natural Dun	May-June	June-July	11:00 a.m. — 3:00 p.m.
Salmon Fly	Nymph: Box Canyon Stone	May-June	June-July	Night
	Adult: Langtry Special			
Golden Stone	Stimulator	June	June-July	Night
Pale Morning Dun	Emerger: Emerging PMD	May-July	June-October	9:00 a.m. — 3:00 p.m.
	Dun: Pale Morning Comparadun			
Green Damsel	Nymph: Green Damsel	May-July	May-August	4:00 p.m. — 9:00 p.m.
Gray Sedge	Larva: Green Caddis Larva	May-July	June-August	11:00 a.m. — 9:00 p.m.
	Pupa: Partridge and Green and Fur Thorax			
	Adult: Deer Hair Caddis			

HATCH CHART
Emergence Period

Natural	Artificial	Coast	Rockies	Emergence Time
Spotted Sedge	Larva: Tan Caddis Larva	June-September	June-September	11:00 a.m. — 9:00 p.m.
	Pupa: March Brown Spider			
	Adult: Elk Hair Caddis			
Grasshoppers	Letort Hopper	July-September	July-September	10:00 a.m. — 7:00 p.m.
Ants	Flying Black Ant	May-October	May-October	10:00 a.m. — 7:00 p.m.
Beetles	Black Crowe Beetle	May-October	May-October	10:00 a.m. — 7:00 p.m.
Tricos	Trico Poly-Wing Spinner	June-September	June-September	7:30 a.m. — 11:00 a.m.
Fall Caddis	Fall Caddis	Sept.-Oct.	Sept.-Oct.	3:00 p.m. — 9:00 p.m.
Scuds	Olive Scud	All Year	All Year	All Day
Midges	Pupa: TDC	Feb.-Nov.	March-Nov.	All Day
	Emergers: Griffith's Gnat			

Index